Preface

This book contains the detailed solutions of all the exercises of my book: The Mechanics of Lorentz Transformations. These solutions are generally very detailed and hence they are supposed to provide some sort of revision for the subject topic. I hope these solutions will be useful to the readers and users of my original book. However, I strongly encourage the readers and users to refer to these solutions only after they solve the exercises, or at least after they make a serious effort, so that consolidation and acquiring the necessary skills, which are the objectives of these exercises will not be lost or diminished.
Taha Sochi
London, August 2018

Contents

Preface 1

Table of Contents 2

Nomenclature 7

1 Preliminaries — 10
1.1 General Background 10
1.2 Historical Issues and Credits 11
1.3 General Terminology 12
1.4 Mathematical Preliminaries 14
1.5 General Conventions, Notations and Remarks 16
1.6 Physical Reality and Truth 21
1.7 Intrinsic and Extrinsic Properties 22
1.8 Invariance of Physical Laws 23
1.9 Galilean Transformations 25
    1.9.1 Space and Time Transformations 26
    1.9.2 Velocity Transformations 28
    1.9.3 Velocity Composition 33
    1.9.4 Acceleration and Force Transformations 34
1.10 Newton's Laws of Motion 39
1.11 Thought Experiments 42
1.12 Requirements for Scientific Theories and Facts 43
1.13 Speed 44
    1.13.1 Speed of Projectile 44
    1.13.2 Speed of Wave 44
1.14 Speed of Light 46

2 Emergence of Lorentz Mechanics — 48
2.1 Classical View of the World 48
2.2 Galilean Relativity 52
2.3 Maxwell's Equations and Speed of Light 55
2.4 Maxwell's Equations and Galilean Transformations 56
2.5 Light as Wave Phenomenon and Luminiferous Ether 58
2.6 Michelson-Morley Experiment 61
2.7 FitzGerald-Lorentz Proposal and Emergence of Lorentz Transformations 63
2.8 Poincare Suggestion and Subsequent Developments 66

3 Introduction to Lorentz Mechanics   69
- 3.1 Lorentz Mechanics versus Other Mechanics 69
- 3.2 Restrictions and Conditions on Lorentz Mechanics 70
- 3.3 Space Coordination 72
- 3.4 Time Measurement and Synchronization 73
- 3.5 Calibration of Space and Time Measurement 77
- 3.6 Reference Frame 77
  - 3.6.1 Construction of Reference Frame 77
  - 3.6.2 Inertial and non-Inertial Frames 79
  - 3.6.3 Difference between Inertial and non-Inertial Frames 83
- 3.7 Causal Relations 84
- 3.8 Speed of Light 85
  - 3.8.1 Speed of Light as Restricted and Ultimate Speed 86
  - 3.8.2 Measuring the Speed of Light 88
- 3.9 Spacetime in Lorentz Mechanics 89
  - 3.9.1 Spacetime Diagram and World Line 90
  - 3.9.2 Light Cone 92
  - 3.9.3 Spacetime Interval 96
  - 3.9.4 Invariance of Spacetime Interval 101

4 Formalism of Lorentz Mechanics   105
- 4.1 Physical Quantities 105
  - 4.1.1 Length 105
  - 4.1.2 Time Interval 109
  - 4.1.3 Mass 113
  - 4.1.4 Velocity 114
  - 4.1.5 Acceleration 116
  - 4.1.6 Momentum 116
  - 4.1.7 Force 117
  - 4.1.8 Energy 118
  - 4.1.9 Work 122
- 4.2 Physical Transformations 123
  - 4.2.1 Lorentz Spacetime Coordinate Transformations 124
  - 4.2.2 Velocity Transformations 128
  - 4.2.3 Velocity Composition 134
  - 4.2.4 Acceleration Transformations 140
  - 4.2.5 Length Transformation 141
  - 4.2.6 Time Interval Transformation 141
  - 4.2.7 Mass Transformation 141
  - 4.2.8 Frequency Transformation and Doppler Shift 141
  - 4.2.9 Charge Density and Current Density Transformations 146
- 4.3 Physical Relations 147
  - 4.3.1 Newton's Second Law 147

		4.3.2 Mass-Energy Relation	147

        4.3.2 Mass-Energy Relation 147
        4.3.3 Momentum-Energy Relation 151
        4.3.4 Work-Energy Relation 154
  4.4 Conservation Laws 154
  4.5 Restoring Classical Formulation at Low Speed 157
  4.6 Restrictions at High Speed 158

5 Derivation of Formalism   162
  5.1 Physical Transformations 162
        5.1.1 Lorentz Spacetime Coordinate Transformations 163
        5.1.2 Velocity Transformations 164
        5.1.3 Acceleration Transformations 164
        5.1.4 Length Transformation 166
        5.1.5 Time Interval Transformation 166
        5.1.6 Mass Transformation 168
        5.1.7 Frequency Transformation and Doppler Shift 169
        5.1.8 Charge Density and Current Density Transformations 169
  5.2 Physical Quantities 169
        5.2.1 Momentum 169
        5.2.2 Force 171
        5.2.3 Energy 171
  5.3 Physical Relations 172
        5.3.1 Newton's Second Law 172
        5.3.2 Mass-Energy Relation 173
        5.3.3 Momentum-Energy Relation 175
        5.3.4 Work-Energy Relation 175
  5.4 Conservation Laws 175

6 Tensor Formulation of Lorentz Mechanics   178
  6.1 Preliminaries 178
  6.2 Useful Mathematics 179
  6.3 Minkowski Metric Tensor 180
  6.4 Lorentz Transformations in Matrix and Tensor Form 183
  6.5 Vector, Tensor and Matrix Formulation 183
        6.5.1 Spacetime Position and Displacement 4-Vector 183
        6.5.2 Quadratic Form of Spacetime Interval 183
        6.5.3 Velocity 184
        6.5.4 Acceleration 184
        6.5.5 Momentum 185
        6.5.6 Force and Newton's Second Law 187
        6.5.7 Electromagnetism and Maxwell's Equations 188

7 Consequences and Predictions of Lorentz Mechanics — 190
7.1 Merging of Space and Time into Spacetime — 190
7.2 Length Contraction — 190
7.3 Time Dilation — 190
7.4 Relativity of Simultaneity — 191
7.5 Relativity of Co-positionality — 195
7.6 Equivalence of Mass and Energy — 197

8 Evidence for Lorentz Mechanics — 198
8.1 Success of Lorentz Transformations — 198
8.2 Mass-Energy Equivalence — 198
8.3 Prolongation of Lifetime of Elementary Particles — 198
8.4 Atomic Clock Experiment — 200
8.5 Stellar Aberration — 201

9 Special Relativity — 202
9.1 Characteristic Features of Special Relativity — 202
9.2 Postulates of Special Relativity — 203
9.3 Assessing the Postulates of Special Relativity — 204
9.3.1 Relativity Principle — 204
9.3.2 Invariance of Observed Speed of Light — 206
9.3.3 Overall Assessment of Special Relativity Postulates — 209
9.4 Abolishment of Fundamental Concepts — 210
9.5 Controversies within Special Relativity — 211
9.6 Thought Experiments in Special Relativity — 212
9.6.1 Train Thought Experiment — 212
9.7 Light Clock — 216
9.7.1 Assessing Light Clock — 221

10 Challenges, Criticisms and Controversies — 223
10.1 Twin Paradox — 223
10.1.1 Time Dilation Effect is Apparent — 224
10.1.2 Traveling Twin is Distinguished by being non-Inertial — 224
10.1.3 Traveling Twin has Two Inertial Frames — 224
10.1.4 Calling for General Relativity — 225
10.2 Barn-Pole Paradox — 225
10.3 Other Paradoxes — 227
10.4 Speeds Exceeding c — 227
10.5 Non-Local Reality of Quantum Mechanics — 227

11 Interpretation of Lorentz Mechanics — 228
11.1 Criteria for Acceptable Interpretation — 229
11.2 Essential Elements of Potentially Acceptable Interpretation — 231

12 Appendices   239

12.1 Maxwell's Equations . . . 239
12.2 Michelson-Morley Experiment . . . 239
12.3 Invariance of Laws under Galilean and Lorentz Transformations . . . 239
12.3.1 Laws of Classical Mechanics . . . 240
12.3.2 Maxwell's Equations . . . 240
12.3.3 Electromagnetic Wave Equation . . . 240
12.4 Derivation of Lorentz Spacetime Coordinate Transformations . . . 240
12.4.1 Special Relativity Method of Derivation . . . 240
12.4.2 Our Method of Derivation . . . 240

Index   241

Nomenclature

In the following list, we define the common symbols, notations and abbreviations that are used in the book as a quick reference for the reader.

∇	nabla differential operator in ordinary 3D space
∇h	gradient of scalar h
$\nabla \cdot \mathbf{A}$	divergence of vector $\mathbf{A}$
$\nabla \times \mathbf{A}$	curl of vector $\mathbf{A}$
∇^2	Laplacian operator in ordinary 3D space
$\Box$	nabla operator in Minkowski 4D spacetime
$\Box^2$	d'Alembertian operator in Minkowski 4D spacetime
$\partial_\mu, \partial^\mu$	partial derivative with respect to the μth coordinate
$\delta/\delta\tau$	absolute or intrinsic derivative operator with respect to τ
$\sim$	comparable in size
′ (prime)	mark of reference frame in motion relative to a given reference frame
0 (subscript)	proper quantity, e.g. proper length L_0
a	magnitude of acceleration or 1D acceleration
$\mathbf{a}, \mathbf{A}$	acceleration vector in 3D, 4D
$\mathbf{a}, \mathbf{A}$	electromagnetic vector potential in 3D, 4D
a^i, A^μ	acceleration vector or its components in 3D, 4D
a^i, A^μ	electromagnetic vector potential or its components in 3D, 4D
a_x, a_y, a_z	components of 3D acceleration vector in x, y, z directions
a_x, a_y, a_z	components of 3D electromagnetic vector potential in x, y, z directions
B	magnitude of magnetic field
$\mathbf{B}$	magnetic field vector
B_x, B_y, B_z	components of magnetic field vector in x, y, z directions
c	characteristic speed of light in vacuum
$\text{diag}[\cdots]$	diagonal matrix with embraced diagonal elements
$ds, d\sigma$	infinitesimal line element in 3D space, 4D spacetime
E	energy
$\mathbf{E}$	electric field vector
E_0, E_k, E_t	rest, kinetic, total energy
E_x, E_y, E_z	components of electric field vector in x, y, z directions
Eq./Eqs.	Equation/Equations
f	magnitude of force or 1D force
$\mathbf{f}, \mathbf{F}$	force vector in 3D, 4D
f^i, F^μ	3D, 4D force vector or its components
f_x, f_y, f_z	components of 3D force vector in x, y, z directions
g_{ij}, g^{ij}	covariant, contravariant metric tensor of 3D space or its components
$g_{\mu\nu}, g^{\mu\nu}$	covariant, contravariant metric tensor of 4D spacetime or its components
i	imaginary unit

iff	if and only if
j, **J**	electric current density vector in 3D, 4D
j_x, j_y, j_z	components of 3D electric current density vector in x, y, z directions
J^μ	electric current density vector or its components in 4D
k	dragging coefficient
k_w	dragging coefficient of water
L, L_0	length, proper length
L, $[L^\mu_\nu]$	Lorentz matrix
$\mathbf{L}^{-1}$	inverse of Lorentz matrix
L^μ_ν	Lorentz tensor or its components
Ly	light year (distance)
m, m_0	mass, rest mass
m_e, m_n, m_p	mass of electron, neutron, proton
M^μ_ν	inverse of Lorentz tensor or its components
n	refractive index
nD	n-dimensional
O	observer or frame of reference
p	magnitude of momentum or 1D momentum
p, **P**	3D, 4D momentum vector
p^i, P^μ	3D, 4D momentum vector or its components
p_x, p_y, p_z	components of 3D momentum vector in x, y, z directions
Q	electric charge
r	radius
r	position vector in 3D space, i.e (x, y, z) or (x^1, x^2, x^3)
r, θ, ϕ	spherical coordinates of 3D space
s	3D space interval (or length of arc)
S, $[S^{\mu\nu}]$	electromagnetic field strength matrix
$S^{\mu\nu}$	electromagnetic field strength tensor or its components
t, t_0	time, proper time
T, $[T^{\mu\nu}]$	electromagnetic dual field strength matrix
$T^{\mu\nu}$	electromagnetic dual field strength tensor or its components
u	speed or 1D velocity (usually belongs to observed object)
u, **U**	3D, 4D velocity vector
u^i, U^μ	3D, 4D velocity vector or its components
u_x, u_y, u_z	components of 3D velocity vector in x, y, z directions
v	speed or 1D velocity (usually belongs to inertial frame)
v	3D velocity vector
v_x, v_y, v_z	components of 3D velocity vector in x, y, z directions
V	event
V, V_0	volume, proper volume
x	position vector in 4D spacetime, i.e. (x^0, x^1, x^2, x^3) or (x^1, x^2, x^3, x^4)
x^0, x^1, x^2, x^3	spacetime coordinates (x^0 temporal)
x^1, x^2, x^3, x^4	spacetime coordinates (x^4 temporal)

x^i, x^μ	coordinates of 3D space, 4D spacetime
x, y, z	spatial coordinates (normally rectangular Cartesian)
β	speed ratio
γ	Lorentz factor
$\Gamma^\mu_{\nu\omega}$	Christoffel symbol of 2nd kind for 4D spacetime
δ^μ_ν	Kronecker delta tensor in 4D
Δ	finite change
ε_0	permittivity of free space
λ, λ_0	wavelength, proper wavelength
μ_0	permeability of free space
ν, ν_0	frequency, proper frequency
ρ, ρ_0	charge density, proper charge density
ρ, ϕ, z	cylindrical coordinates of 3D space
σ	spacetime interval in 4D
τ	proper time parameter
ϕ	electromagnetic scalar potential
ω	angular speed

Chapter 1
Preliminaries

1.1 General Background

1. Define the science of mechanics in a few words.
   Answer: The science of mechanics is a branch of physics dedicated to the description and prediction (quantitatively and qualitatively) of the motion of physical objects in space and time and identifying its causes.
2. Make a brief comparison between the mechanics of Lorentz transformations and the following branches and theories of mechanics: Newtonian mechanics, quantum mechanics and general relativistic mechanics.
   Answer:
   Newtonian versus Lorentz mechanics: the focus of Newtonian mechanics is the macroscopic world at normal speeds, while the focus of Lorentz mechanics is objects moving at any physically-possible speed where both these types of mechanics record their observations from inertial frames. Accordingly, Newtonian mechanics is a valid approximation to Lorentz mechanics at low speeds ($v \ll c$), but it must be replaced by Lorentz mechanics at high speeds ($v \sim c$).
   Quantum versus Lorentz mechanics: quantum mechanics is mainly concerned with tiny objects at atomic and subatomic levels whether they are stationary or at low or high speeds, while Lorentz mechanics is mainly concerned with objects moving at high speeds whether they are big or tiny. Hence the two have common interest in tiny objects that move at high speeds.[1]
   General relativity versus Lorentz mechanics: general relativity is a theory of gravity while Lorentz mechanics is a theory about the laws of physics as seen from inertial frames of reference. In fact, the essence of Lorentz mechanics is space and time and how they are transformed between different inertial frames, as we will see later.
3. Why we think it is better to use expressions like "mechanics of Lorentz transformations" or "Lorentz mechanics" instead of the common expressions like "special relativity" or "special relativistic mechanics" to refer to the part of mechanics that is based on the Lorentz transformations?
   Answer: Because the preferred expressions are more inclusive and objective since they are based on the bare formalism with no association or adoption of a particular epistemological or philosophical framework, while the common expressions associate this branch of mechanics with the special theory of relativity which essentially represents a particular interpretation of Lorentz mechanics, and hence its characteristic epistemological and philosophical framework excludes other potential interpretations and may entail unverified or unverifiable implications which can lead to wrong conclusions. Another reason is related to the credit of creating and developing this branch of physics

---
[1] The two types of mechanics are merged into the so-called relativistic quantum mechanics.

1.2 Historical Issues and Credits 11

where we believe that many scientists contributed to the emergence and rise of Lorentz mechanics in contrast to the common belief that this branch of modern physics belongs to Einstein and his contribution to the development of special relativity theory of Poincare. This strong association between Einstein and special relativity theory makes the use of the common expressions with the distinctive "relativistic" attribute susceptible to the wrong interpretation or impression that this branch of mechanics is the brainchild of Einstein.

4. Compare between the approach followed in the construction and presentation of quantum mechanics in the common textbooks of physics and the approach followed in the construction and presentation of Lorentz mechanics in these texts to see if the two subjects are treated equally.
Answer: While the construction and development of quantum mechanics in the textbooks start from the formalism and end with the interpretations, the construction and development of Lorentz mechanics (which is presented as special relativity) start from the interpretation (represented by the postulates of special relativity and their consequences) and end with the formalism where the formalism and all its consequences are supposed to be completely based on this particular interpretation. Hence, the two subjects are treated differently with no justification and this can be regarded as bias in favor of the special relativistic view.

1.2 Historical Issues and Credits

1. Name a number of physicists who have contributed to the development of Lorentz mechanics in its early days.
   Answer: Some of these physicists are: FitzGerald, Lorentz, Larmor, Voigt, Poincare, Einstein, Planck, Laue, Minkowski and Sommerfeld.
2. When the credit for the creation and development of Lorentz mechanics started to shift from Lorentz, Poincare and other major contributors to Einstein?
   Answer: This started after the alleged confirmation of general relativity by the solar eclipse expedition of Eddington and his team in 1919. What followed of public euphoria and the huge inflation in the image of Einstein led to many fabrications and falsifications including this one where Lorentz mechanics became the brainchild of Einstein. Before that, this part of physics was generally associated with the work and views of Lorentz and Poincare as well as many other primary and secondary contributors.
3. Why we attribute the main credit for the special relativity interpretation to Poincare?
   Answer: Because all the main elements of the modern theory of special relativity belong to Poincare;[2] examples of these are:
   • Questioning the existence (or at least the necessity) of ether and hence originating the principle of relativity in its special relativistic sense.
   • Inventing the synchronization procedure of clocks which is based on the use of light signals. This procedure is based in a sense on the universality of the speed of light (see

---

[2] However, the Poincare version is apparently more rational and consistent than the modern version of special relativity which generally reflects Einstein views.

§ 3.4 in the main text and exercises).
• Questioning the difference between time and local time which eventually led to the proposal of the abolishment of absolute time. Also, time dilation as an interpretation to the difference in the time measurements between frames is attributed to Poincare.[3]
• Using the concept of 4D spacetime as the space of Lorentz mechanics. However, there are indications that this idea was in circulation even before Poincare.
• The requirement of the physical laws in general (and not only the laws of mechanics) to be invariant under the correct set of coordinate transformations which is the essence of the relativity postulate of special relativity. A proof of the invariance of Maxwell's equations (as an example of this general invariance principle) under the Lorentz transformations is also attributed to him (among others like Lorentz and Larmor).

Apart from all these contributions, as well as many other contributions, all the elements of special relativity were hotly debated issues before the work of Einstein and hence almost all of Einstein work was reflection of what was already there in the scientific circles and journals with some elaborations and extensions here and there. So, even some of the elements of special relativity which cannot be firmly traced to Poincare can be considered as part of the common literature of this branch of physics at that time. Accordingly, Einstein was a secondary contributor not only to the formalism of Lorentz mechanics but even to its special relativity interpretation and therefore we believe that the main credit for special relativity theory belongs to Poincare and his original ideas not to Einstein.

1.3 General Terminology

1. Briefly define the following terms: massive object, massless object, uniform motion, inertial frame, rest frame, proper quantity, Minkowski space, world line, event, simultaneous events, co-positional events, space interval, and spacetime interval.
Answer:
Massive object is an object with non-vanishing mass (i.e. $m > 0$) like proton.
Massless object is an object without mass (i.e. $m = 0$) like photon.
Uniform motion is a motion with constant velocity, i.e. constant speed in a given direction so both the magnitude and the direction of the velocity are independent of time.
Inertial frame is a frame of reference in which Newton's three laws of motion hold true.
Rest frame of an object is a frame of reference in which the object is at rest and not moving.
Proper quantity is the quantity of an object (e.g. the length of a stick) as measured in its rest frame.
Minkowski space (or spacetime) is a 4D manifold consisting of one temporal dimension and three spatial dimensions.
World line is the path traced in the Minkowski spacetime by an object and hence it represents its presence in spacetime or alternatively its trajectory or footprint in this

---
[3] We note that length contraction, which was originally proposed by FitzGerald and adopted by Lorentz, was already in existence as an explanation for the null result of Michelson-Morley experiment.

4D manifold. World line may also represent a continuous series of correlated events.
Event is a physical occurrence that takes place in the spacetime; it is usually represented by a point in the spacetime.
Simultaneous events are events that take place at the same time.
Co-positional events are events that take place in the same location of space.
Space interval is the length of a segment in the ordinary (or spatial) 3D space.
Spacetime interval is the length (but it can be imaginary!) of a segment in the 4D spacetime manifold.

2. What is the difference between coordinate system and frame of reference?
Answer: Coordinate system is an abstract spatial device used to locate points in space, while frame of reference is an abstract device used to identify points in spacetime manifold and hence it is a coordinate system associated with a time measuring mechanism. So, frame of reference is a "coordinate system" of spacetime, while coordinate system is a "coordinate system" of space.

3. What is the difference between "space and time" and "spacetime"?
Answer: The difference is that "space and time" is based on the distinction between the spatial and temporal dimensions which is the view of classical mechanics, while "spacetime" is based on the entanglement of space and time into a single entity or manifold where the temporal and spatial dimensions are treated equally and this is the view of Lorentz mechanics.

4. List a number of transformations between different frames of reference which are at rest relative to each other.
Answer: The following are some examples:
• Scaling of the units of space (length) or/and time.
• Static translation of origin of coordinates or/and origin of time.
• Static rotation of coordinate system in space.
• Shearing of space coordinate axes.
• Reflection of space coordinate axes. Reflection of time axis (i.e. in the same sense as reflection of space coordinate axes) is apparently not considered in the common literature of physics due to the explicit or implicit assumption of the uniqueness of the direction of time flow (i.e. past→present→future) although a sign difference (of only symbolic significance) between the time of two frames is possible to consider.

5. List a number of differences between different reference frames which are in a state of motion relative to each other.
Answer: As well as some of the differences that originate from the transformations in the previous exercise (e.g. scaling of space and time units), the following can also take place:
• Being in a state of relative uniform translational motion.
• Being in a state of relative uniform rotational motion.
• Being in a state of relative uniform acceleration (whether translational, rotational or both).
• Being in a state of relative non-uniform acceleration (whether translational, rotational or both).

We note that some combinations of the above states (e.g. relative uniform translational motion with relative uniform rotational acceleration) are also possible.

6. Which of the differences that we considered in the previous two exercises do not affect the status of inertiality (i.e. the state of being inertial or non-inertial) between two reference frames?
   Answer: It is obvious that all the differences in exercise 4 do not affect the status of inertiality of two frames because the two frames are assumed to be in a state of relative rest and hence they both should be inertial or non-inertial as the two frames are essentially identical apart from possible static transformations which do not affect their inertiality status that is based on motion and dynamic transformations. Regarding the differences in exercise 5, being in a state of relative uniform translational motion does not affect the status of inertiality (i.e. the two frames are either both inertial or both non-inertial) but the other three differences could affect the status of inertiality between two frames. In brief, being in a state of relative uniform rotational motion or uniform acceleration or non-uniform acceleration can take place between two non-inertial frames as well as between an inertial frame and a non-inertial frame. More details about this issue will come later in the book (see for example § 3.6.2).

1.4 Mathematical Preliminaries

1. Define β and γ as used in the literature of Lorentz mechanics.
   Answer: The speed ratio β and the Lorentz factor γ are defined in the literature of Lorentz mechanics by the following expressions:
   $$\beta = \frac{v}{c} \qquad \gamma = \frac{1}{\sqrt{1-\beta^2}}$$
   where v is the speed of an object or a frame of reference and c is the characteristic speed of light in free space.

2. Show that:
   $$\gamma^2 = \gamma^2 \beta^2 + 1$$
   Answer: This can be obtained directly from the identity $\gamma^2 - 1 = \gamma^2 \beta^2$ or the identity $\gamma^2 - \gamma^2 \beta^2 = 1$. It can also be obtained from scratch as follows:
   $$\begin{aligned}
   \gamma^2 &= \frac{1}{1-\beta^2} \\
   &= \frac{1+\beta^2-\beta^2}{1-\beta^2} \\
   &= \frac{\beta^2}{1-\beta^2} + \frac{1-\beta^2}{1-\beta^2} \\
   &= \frac{\beta^2}{1-\beta^2} + 1 \\
   &= \gamma^2 \beta^2 + 1
   \end{aligned}$$

1.4 Mathematical Preliminaries

3. Show that:
$$\gamma + 1 = \frac{\gamma^2 \beta^2}{\gamma - 1} \qquad (\gamma \neq 1)$$
   Answer: This is just the identity $\gamma^2 - 1 = \gamma^2 \beta^2$ in a disguised from, that is:
$$\begin{aligned} \gamma^2 - 1 &= \gamma^2 \beta^2 \\ (\gamma + 1)(\gamma - 1) &= \gamma^2 \beta^2 \\ \gamma + 1 &= \frac{\gamma^2 \beta^2}{\gamma - 1} \end{aligned}$$

4. Find the relative error in using the approximation:
$$\gamma \simeq 1 + \frac{1}{2}\beta^2$$
   when $v = 0.1c$ and hence assess the reliability of this approximation.
   Answer: The relative error can be calculated from the expression:
$$\frac{\gamma - \left(1 + \frac{1}{2}\beta^2\right)}{\gamma}$$
   which is about 3.76×10^{-5} when $v = 0.1c$. This error is very small although the speed $v = 0.1c$ is very high by classical standards. Hence, the above approximation is reliable when $v \ll c$.

5. Repeat the previous question with the approximation:
$$\gamma^2 - \gamma \simeq \frac{\beta^2}{2}$$
   Answer: The relative error can be calculated from the expression:
$$\frac{(\gamma^2 - \gamma) - \frac{\beta^2}{2}}{\gamma^2 - \gamma}$$
   which is about 0.0125 (i.e. about 1.25%) when $v = 0.1c$ which is classically very high speed. This error is negligible in many practical situations and hence the above approximation is fairly reliable when $v \ll c$.

6. Plot the Lorentz γ factor as a function of the speed ratio β and discuss the distinct features of this plot and the physical significance of these features on the relation between classical and Lorentz mechanics and on the issue of speed restrictions in Lorentz mechanics.
   Answer: The plot should look more or less like Figure 1. The following features can be easily observed:
   • At $v = 0$ (and hence $\beta = 0$), $\gamma = 1$. In this case, the formalism of classical mechanics is exact since the observed frame or object is at rest with respect to the observer.
   • At low speeds (i.e. $v \ll c$ and hence $\beta \simeq 0$) γ stays very close to unity. In this

case, the formalism of classical mechanics is a good approximation to the formalism of Lorentz mechanics.
- At very high speeds (i.e. $v \simeq c$ and hence $\beta \simeq 1$) γ shoots up sharply where the line $\beta = 1$ is a vertical asymptote to the γ curve. In this case, the formalism of Lorentz mechanics must replace the formalism of classical mechanics.
- The point $\beta = 1$ is a singularity point for γ and hence the formalism of Lorentz mechanics fails to provide any prediction about this case. Similarly, the case $\beta > 1$ makes γ imaginary and hence it is physically meaningless. These cases (i.e. $\beta \geq 1$) are generally interpreted as an implication that reaching or exceeding the speed of light is physically impossible. However, we would rather say it represents a limitation of Lorentz mechanics in its current formulation and hence although these cases are banned by the current formulation of Lorentz mechanics, they may be possible under an amendment or an extension of the current formulation or under a more comprehensive mechanics where Lorentz mechanics plays the role of an approximation or a limiting case for this mechanics, like classical mechanics in its relation to Lorentz mechanics. We should also point out to the fact that because Lorentz mechanics is restricted to inertial frames, reaching or exceeding the characteristic speed of light c will not be banned automatically outside the domain of Lorentz mechanics (i.e. inertial frames) because of its ban in Lorentz mechanics. Hence, any claim of banning these cases outside the domain of Lorentz mechanics requires a new and independent evidence other than the singularity of γ or being imaginary when $v \geq c$. We also note that there should be a distinction in the speed restriction $v < c$ between massive and massless objects as will be discussed later. In this context we may say, even if the formalism of Lorentz mechanics implies that c is a restricted and ultimate speed to light (including all types of electromagnetic radiation), it does not automatically extend to massless physical objects other than light, i.e. we need an independent proof for this premise other than the formalism of Lorentz mechanics.

1.5 General Conventions, Notations and Remarks

1. Why "light" in the literature of Lorentz mechanics includes all types of electromagnetic radiation?
   Answer: Because what is important in Lorentz mechanics is the speed of propagation in free space which is the same for all types of electromagnetic radiation and hence they are all equal in this regard. As we will see, this speed may also be extended to include all massless objects according to special relativity.
2. Discuss the difference between the *characteristic* and *observed* speed of light in free space.
   Answer: We should clearly distinguish between the constant c ($\simeq 3 \times 10^8$ m/s) which symbolizes the characteristic speed of light, and the observed speed of light which is the actually observed speed of light and hence in principle it may or may not be equal to c. For example, if the observed speed of light is alleged to be frame dependent then this speed may be given as $c \pm v$ where the constant c represents the characteristic speed of

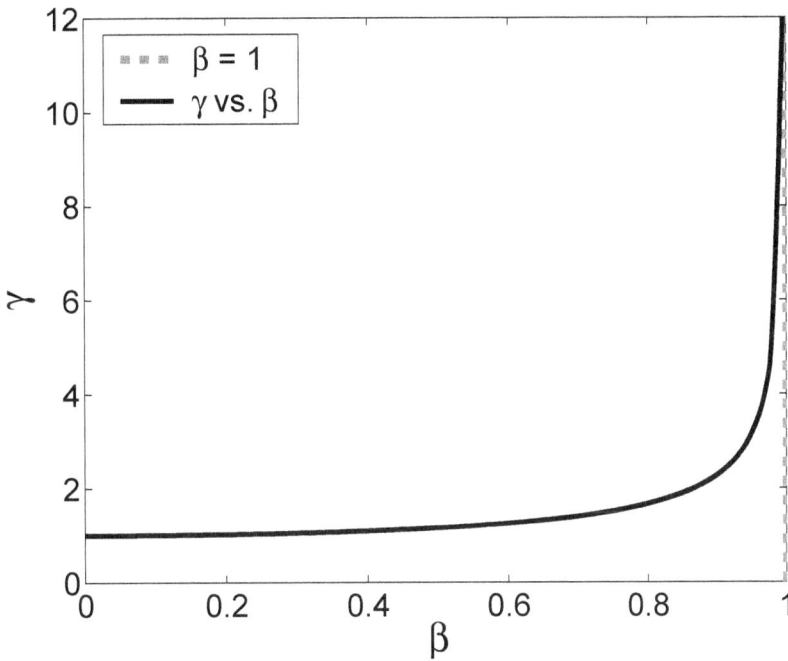

Figure 1: The Lorentz γ factor as a function of the speed ratio β.

light while v is the frame dependent part of the observed speed. Although we generally distinguish the two by the "characteristic" and "observed" labels, in some cases we may use "the speed of light" without these labels where the meaning can be judged from the context. Based on this distinction, the reader should not be confused to believe that the constancy of c means the acceptance of the second postulate of special relativity whose essence is the constancy of the observed (not the characteristic) speed of light.[4]

3. Expand the following expressions using the adopted conventions in this book:
$$u_i u^i \qquad cA^\mu b_\mu \qquad \mathbf{r} \qquad \mathbf{x}$$

Answer: Using the summation and index range conventions as well as other symbolic conventions, we obtain:

$$u_i u^i \equiv \sum_{i=1}^{3} u_i u^i = u_1 u^1 + u_2 u^2 + u_3 u^3$$

$$cA^\mu b_\mu \equiv \sum_{\mu=1}^{4} cA^\mu b_\mu = cA^1 b_1 + cA^2 b_2 + cA^3 b_3 + cA^4 b_4$$

$$\mathbf{r} = \left(x^1, x^2, x^3\right) = (x, y, z)$$

$$\mathbf{x} = \left(x^1, x^2, x^3, x^4\right) = (x, y, z, ct)$$

---

[4] In brief, the characteristic speed of light c is the same for all observers while the observed speed of light is *in principle* observer-dependent. Hence, the essence of the second postulate of special relativity is the claim that the observed speed of light is always equal to the characteristic speed of light c and hence the observed speed of light is observer-independent.

1.5 General Conventions, Notations and Remarks

The indices in the second and fourth examples may also range over $0, 1, 2, 3$ instead of $1, 2, 3, 4$ and hence the order of the spatial and temporal variables will change accordingly.

4. Explain the standard setting between two inertial frames illustrating your descriptive explanation by a simple sketch.
Answer: The main features of the standard setting between two inertial frames of reference, O and O', which are in a state of relative uniform translational motion are outlined in the following bullet points:
• The two frames have rectangular Cartesian coordinate systems with common length unit and common time unit.
• These coordinate systems become identical at time $t = t' = 0$ where the axes have the same sense of orientation, i.e. the positive and negative directions of the corresponding axes (i.e. x-x', y-y' and z-z') are identical. Accordingly, the origins of coordinate of the two systems are identical at time $t = t' = 0$. It should be obvious that these frames have identical origins of time, i.e. $t = t' = 0$, which coincide when their origins of spatial coordinates coincide.
• The two frames are in a state of relative translational motion with velocity v along the common x-x' direction only and hence they are at relative rest in the y-y' and z-z' directions. Accordingly, the y-y' and z-z' coordinates of any event or object seen from these frames remain identical at all times. We note that the symbol v stands for the velocity of O' relative to O and hence v is positive/negative if O' is seen by O to move in the positive/negative direction of the common x-x' axis.
• Since the two frames are assumed to be inertial, the relative motion between these frames is purely translational and hence there are no other forms of motion like rotational, whether uniform or non-uniform. In fact, for the two frames to be inertial there should be no rotational motion whether relative or not, as indicated in the text.
• The velocity of the relative motion along the common x-x' direction is constant and hence the speed and direction of this motion are independent of time.
The required illustrative sketch should look like Figure 2.

5. What are the main advantages of using the standard setting?
Answer: The main advantages can be summarized in the following points:
• The two coordinate systems are chosen to be rectangular Cartesian, which is the simplest form of coordinate systems, and this results in considerable simplification in formulation.
• The origins of coordinates are made to coincide at time $t = t' = 0$ and this results in further simplification. This also applies to the unification of their origins of time, i.e. $t = t' = 0$.
• The motion is oriented to be only along the common x-x' axis and hence the motion is made essentially one-dimensional although it takes place in a 3D space. This makes the formulation and notation easier since a scalar-like approach in the formulation and notation can be used instead of a strict vector or tensor approach which is more complex and demanding.
• The two frames are fully distinguished by a single scalar-like number v that represents

1.5 General Conventions, Notations and Remarks

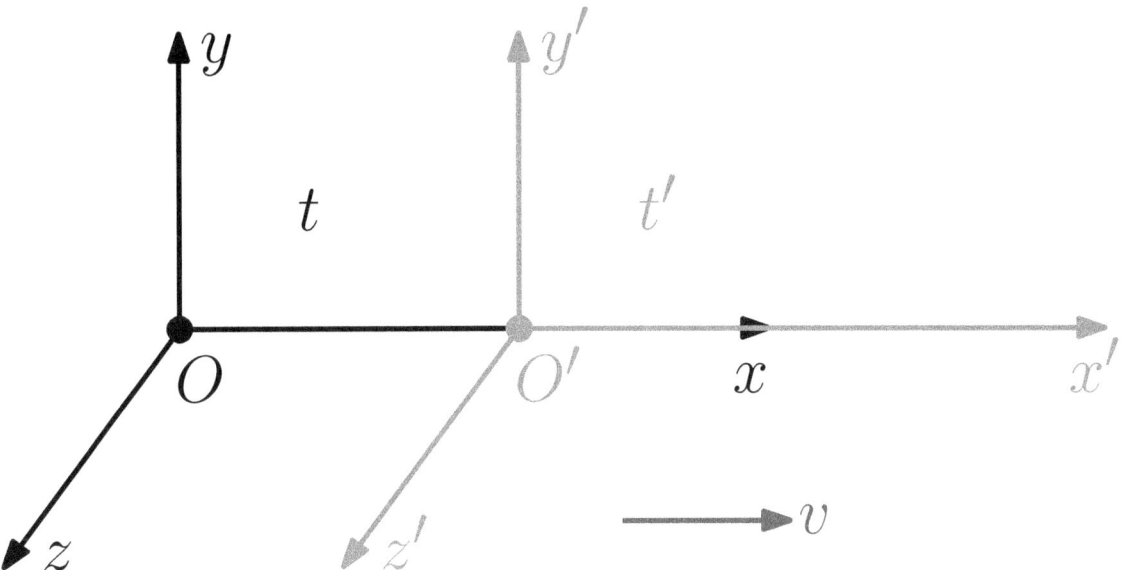

Figure 2: Two inertial frames, O and O', in a state of standard setting.

the relative velocity between the two frames.

6. Justify the common practice in the literature of Lorentz mechanics of using scalar symbols to represent vector quantities like velocity and momentum.
   Answer: This practice is justified by the employment of standard setting between frames (or between frame and observed object and hence it is like standard setting) which essentially reduces most vector quantities to be one-dimensional, i.e. in the x-x' dimension, and hence these quantities can be represented like scalars where the magnitude and sign become sufficient for full identification and representation of these quantities.

7. Why symbols like v may be described as speed sometimes and as velocity in others?
   Answer: As discussed earlier, due to the employment of standard setting between frames (or its alike between frame and object) the motion is one-dimensional, and hence scalar symbols such as v are generally used to represent vectors like the velocity vector $\mathbf{v}$ due to the vanishing of the y and z components. Accordingly, we may describe a symbol like v as speed or as velocity where in the former case we mean the magnitude only while in the latter case we mean the magnitude with sign, i.e. in the positive or negative direction of the line of motion which is the x axis according to the standard setting.

8. In a state of standard setting between two inertial frames: (a) How many numbers associate any given event to fully identify its presence in spacetime? (b) How many of these numbers are independent? (c) How are these numbers related? (d) What is the physical quantity that distinguishes these frames?
   Answer:
   (a) Eight numbers, i.e. (x, y, z, t) that belong to frame O and (x', y', z', t') that belong to frame O'.

(b) Since $y = y'$ and $z = z'$, only six of these numbers are independent.[5]

(c) The two sets of coordinates, i.e. (x, y, z, t) and (x', y', z', t'), are linked by certain transformations (i.e. Galilean in classical mechanics and Lorentzian in Lorentz mechanics) where each one of these sets can be obtained from the other set by applying these transformations.

(d) The physical quantity that distinguishes these frames is the relative velocity v which identifies the speed and direction of their relative motion. The common convention is that v is the velocity of the primed frame as seen from the unprimed frame.

9. Classify the states of two reference frames, whether inertial or non-inertial, in their relation to each other from the perspective of their relative motion. Which of these states represent the same inertiality status of the two frames, i.e. both frames are necessarily inertial or necessarily non-inertial?

   Answer: Broadly speaking, two frames of reference can be in one of the following states:

   (a) In a state of relative rest.
   (b) In a state of relative uniform translational motion.
   (c) In a state of relative uniform rotational motion.
   (d) In a state of relative uniform acceleration (translational or rotational acceleration).
   (e) In a state of relative non-uniform acceleration (translational or rotational acceleration).

   In states (a) and (b), both frames necessarily have the same inertiality status and hence both should be inertial or both should be non-inertial. In states (c), (d) and (e), one of the frames is necessarily non-inertial while the other frame could be inertial or non-inertial and hence they do not necessarily share the same inertiality status.

   We note that the above cases are the principal cases; other cases can be obtained from certain combinations of the above cases as well as from considering multiple directions. The results should be easy to analyze based on the above classification.

10. Give an example of the tendency in the scientific circles to support certain theories and individuals which may cast a shadow on the validity of the claimed evidence in support of Lorentz mechanics especially in its relativistic version.

    Answer: A prominent example of this sort of tendency and prejudice, although it is not related to Lorentz mechanics directly, is the 1919 astronomical expedition where members of the expedition team sexed up the results and hence they declared the "verification" of general relativity predictions which is inconsistent with the actual evidence. Therefore, more caution and skepticism are required in examining and accepting any claimed evidence or support in these emotionally-sensitive areas of physics.

---

[5] In fact, this applies to Lorentz mechanics where $t \ne t'$ in general, but in classical mechanics where $t = t'$ we should have only five independent numbers although this is not because of the standard setting alone but because of the Galilean transformations as well, i.e. the standard setting unifies the origin of time while the Galilean transformations unify the time flow (or time intervals) in the two frames.

1.6 Physical Reality and Truth

1. What are the three principles of reality and truth which are embedded in realistic philosophies and sciences?
   Answer: They are:
   (a) There exists a real world which is independent of the observer.
   (b) The reality of this world is unique.
   (c) The truth, which is an accurate reflection of the unique reality, is unique so any individual physical reality is represented by one and only one truth.

2. Briefly discuss realism in the old and modern physics.
   Answer: In the old or classical physics, mainly prior to the emergence of Lorentz mechanics and quantum physics, realism was the dominant view where all branches of physics are based on the principles of reality and truth, or the realism philosophy. However, following the emergence of modern physics, this position became less firm and clear. In fact, it has been challenged even explicitly by some views and interpretations. This is because old physics is based on the view that the observer is just a receptor of the reflection of reality in the observation process while certain branches of modern physics are based on the view that the observation process is an interaction between the observed phenomenon and the observer and hence the observer can affect the observed phenomenon or this phenomenon can depend on the frame of observation.

3. Discuss briefly the general stand of quantum mechanics and Lorentz mechanics about physical reality and truth.
   Answer: Some interpretations of quantum mechanics are not totally objective and compliant with the principles of physical reality and truth because according to these interpretations the presence of the observer can have an impact on the outcome of the experiment and may even determine the nature of the observed phenomenon. Moreover, we may even have multiple realities such as parallel worlds. However, this form of non-objectivity and violation of the principles of reality and truth should not affect Lorentz mechanics which is based on different philosophical and epistemological framework where the presence of the observer is not supposed to affect the outcome of the experiment, at least in the sense of quantum mechanics, thanks to the total objectivity of physical reality, determinism and preciseness as well as other embedded and generally accepted principles of physical reality and truth which Lorentz mechanics is generally supposed to be based upon. Nonetheless, Lorentz mechanics can also embrace (mostly unwittingly) non-realism views and interpretations where we could have multiple truths, and even multiple realities, that depend on the frame of observation. For example, according to some views in special relativity about length contraction and time dilation and if they are real or apparent effects, we may need to abandon some of the principles of reality and truth to have a sensible and consistent interpretation of these views.

4. Discuss, giving some examples, the issue of instinctive corrections that we apply unconsciously in our daily life and the relation of this to the issues of reality, truth, adaptation and biological evolution.

1.7 Intrinsic and Extrinsic Properties

Answer: There are many examples about the issues of reality and truth from our daily life where our instinctive position is to "correct" for an "error" when we observe something in a different state to the "ideal" state that we consider as the honest representation of the existing and unique physical reality. Many of these corrections are even embedded by evolution in our perception system. For example, when we see a particular object from different perspectives and different distances we introduce, instinctively and unconsciously, all the required corrections to "see" it in the right shape and size where these corrections are based on our past experience and an instinctive tendency for consistency. It should be obvious that this instinctive tendency for introducing these corrections is based on the existence and uniqueness of physical reality and the instinctive realization of the necessity for obtaining a unique "correct" reflection of this reality which we call truth. This is just one example of our elaborate physiological methods for adaptation to the physical world where these principles (i.e. existence and uniqueness of reality and uniqueness of truth) help us to adapt successfully by merging all these different and even contradictory images and perceptions which can cause a huge confusion into a single truth which is the correct reflection and honest representation, according to our instinctive heritage, of the existing and unique reality. Any opposite principles about reality and truth will make the task of adaptation extremely difficult and challenging.

5. Briefly examine the claim that modern science is not subject to the same rules as daily life and the necessity of adaptation, and hence modern science may not need to be based on the principles of reality and truth.
Answer: Modern science is not an exception in the adaptation process and hence it should also be based on the same instinctive and evolutionary principles and rules that govern the biologically-motivated daily life experiences. In fact, science as a highly precise and accurate form of human knowledge should have more reasons to embrace these principles of realism even from a purely pragmatic perspective. Accordingly, if physics should be useful and serve its purpose, it should rely to a certain degree on realism to justify its existence and fulfill its objectives. However, it is difficult to impose strict limits on what is legitimate and what is illegitimate in this regard as long as the philosophical and epistemological position of a particular theory is logically consistent. For example, we may tolerate a form of non-realism in the interpretation of a particular scientific theory, but we cannot accept a scientific or philosophical theory that is entirely based on the rejection of the principles of reality and truth or it contains logical fallacy such as being based in one of its parts on the uniqueness of truth while in another part it is based on the non-uniqueness of truth.

1.7 Intrinsic and Extrinsic Properties

1. What is the difference between intrinsic and extrinsic properties of a physical object? Give examples of each.
Answer: An intrinsic property of a physical object is a property that does not depend in its existence and quantification on any consideration of an external observer or

reference frame, while an extrinsic property is a property that has such a dependency. For example, in classical physics mass, length and size of time duration are intrinsic properties since they do not depend on the observer because if a physical object has a mass of 1 kilogram and a length of 1 meter and it lasts for 1 second it will be so for all observers since these properties belong to the object in itself. On the other hand, velocity and momentum are extrinsic properties because a physical object can have different velocities and momenta depending on the observers and their frames of reference. This is also the case with the improper length and size of time duration in Lorentz mechanics where these physical properties depend on the observer due to length contraction and time dilation effects.

2. Is the acceleration of non-inertial frames an intrinsic or an extrinsic property according to classical mechanics?
Answer: It should be an extrinsic property because although acceleration is the same for all inertial observers (see § 1.9.4) and it is supposed to be referred to an absolute reference frame, its existence and quantification depends on the observer and his reference frame especially if we include non-inertial observers and frames of reference. For example, two non-inertial observers can see each other at rest or in a state of uniform motion (i.e. no acceleration). They can also see each other or another inertial or non-inertial observer to be in a state of acceleration with different measurements of the value of acceleration.

3. Are the extrinsic properties examples of multiple reality or multiple truth?
Answer: No. The reader should not confuse between the uniqueness of reality and truth and the uniqueness of perspective which is part of the reflection of reality or truth. For example, the principles of reality and truth will be violated if the kinetic energy relative to the same observer is 10 J and 100 J at the same time, but these principles will not be violated when the kinetic energy relative to one observer is 10 J and the kinetic energy relative to another observer is 100 J because the perspective of the observer is part of the reflected reality and truth. In brief, when we consider all the factors that determine the reality and define the truth (including the perspective in the extrinsic properties for example) then the reality and truth should be unique.

1.8 Invariance of Physical Laws

1. State briefly the principle of invariance of physical laws giving some examples.
Answer: The principle of invariance of physical laws states that an alleged rule of physics cannot be a law unless it takes the same form in all frames of reference or at least in a certain category of frames. For example, in classical mechanics the principle of conservation of mass is a law of physics because all observers agree on the following form:
$$\Delta m = 0$$
i.e. the net change of mass of a closed physical system is zero. Similarly, Newton's first law of motion is a law of physics because all *inertial* observers agree on the following

1.8 Invariance of Physical Laws 24

form:
$$\sum \mathbf{f} = \mathbf{0} \quad \Leftrightarrow \quad \Delta \mathbf{u} = \mathbf{0}$$
i.e. the net external force $\sum \mathbf{f}$ on a massive object vanishes identically *iff* the change of velocity $\Delta \mathbf{u}$ of the object is zero. On the other hand, the alleged "principle of conservation of speed" or "the principle of constancy of acceleration" are not laws of physics because there is no agreement between any general category of observers that:
$$\Delta u = 0 \quad \text{or} \quad \mathbf{a} = \text{constant}$$
although these "principles" may be valid conditional observations for a number of individual observers.

2. Analyze the essence and roots of the principle of invariance of physical laws and discuss its significance.
   Answer: The essence and roots of this principle is that the form of the physical laws should be invariant under certain set of transformations. For example, the laws of classical mechanics take the same form under the Galilean transformations between inertial observers. Similarly, the laws of mechanics and electromagnetism take the same form under the Lorentz transformations between inertial observers.[6] The significance of this principle is that if any alleged law of physics is to be useful and of common practical value it should be sufficiently general to benefit all observers or at least a large category of observers. This cannot be the case if the form of the alleged law depends on the observer and his frame of reference.

3. State briefly the main assumptions and conditions for the application of the principle of invariance of physical laws.
   Answer: At the root of this principle there are several assumptions and conditions, the main of these are:
   • The existence of a set of transformations such as the Galilean and Lorentzian transformations.
   • The existence of a domain for the validity of this set of transformations, e.g. inertial frames for the Galilean and Lorentzian transformations.
   So, in a given domain where certain transformations apply, a physical rule can be considered a law only if it transforms invariantly by these transformations within the given domain.

4. Make a clear distinction between being value-invariant and being form-invariant.
   Answer: Value-invariance means that the quantitative value of a given physical quantity is the same across different frames, while form-invariance means that the form of a given law or rule is the same across different frames. So in brief, value-invariance is about physical quantities like mass and length, while form-invariance is about laws and rules like Newton's second law or the conservation of momentum. For example, if the

---
[6] In fact, the laws of mechanics in their original classical form are not invariant under the Lorentz transformations, but they are made invariant by certain adjustments to some concepts and definitions (momentum in particular).

mass of a given object is measured as 1 kg by a given inertial observer, then according to the value-invariance of mass in classical mechanics all inertial observers should measure the mass of this object as 1 kg. Similarly, if an inertial observer observed the conservation of momentum in his frame then according to the form-invariance of this principle in Lorentz mechanics all inertial observers should agree on the conservation of momentum although they may disagree on the quantitative value of the conserved momentum.

5. Make a clear distinction between being value-invariant and being constant.
   Answer: Being value-invariant does not necessarily imply being constant. So if we have a variable physical quantity that have the same variable value (as a function of certain variables within given conditions) across all frames then it is value-invariant but it is not constant. On the other hand, we may have a physical quantity that is constant in each frame but its value is not the same in these frames (e.g. it has the constant value 10 in the first frame and the constant value 20 in the second frame) and hence it is constant (say as a function of time) in each frame but it is not value-invariant across these frames. We may also have a physical quantity that is value-invariant across all frames and it is constant so it takes the same constant value, say 10 as a function of time, in all frames. Similarly, we may have a physical quantity that is neither value-invariant across frames nor constant and hence it is variable across frames and in each frame.
   We should remark that "being constant" is usually expressed as "conserved" and hence the conservation of total energy, for example, usually means it is constant as a function of spacetime coordinates (and even other variables), i.e. it does not vary in time or space under certain conditions such as absence of potential fields.

6. Assume that a physical quantity is invariant across all inertial frames and it is conserved (i.e. constant) in a particular inertial frame. What should you conclude?
   Answer: It should be conserved across all inertial frames.

1.9 Galilean Transformations

1. In standard setting where v represents the 1D velocity of O' frame relative to O frame along the common x-x' axis, what is the significance of v being positive, negative or zero?
   Answer: Being positive means that O' is seen by O to move in the positive direction of the x axis while being negative means that O' is seen by O to move in the negative direction of the x axis. Being zero means that the two frames are in a state of relative rest and hence they are essentially identical except, possibly, for a static translation along the x axis.

2. Give an example of other conditions that are usually assumed implicitly in the state of standard setting.
   Answer: An example of such conditions is that both frames use the same length and time measuring equipment and mechanisms and the same physical scales and units so that they can make identical measurements and observations.

1.9.1 Space and Time Transformations

1. How are the Galilean transformations of the primed variables obtained from the Galilean transformations of the unprimed variables and vice versa? Justify your answer.
   Answer: Obtaining the transformations in one direction (i.e. primed to unprimed or the other way around) from the transformations in the opposite direction can be done either by algebraic manipulation, by solving these equations for the opposite variables (with possible use of other transformations), or by exchanging the primed and unprimed labels with reversing the sign of relative velocity. The first method is obviously legitimate since algebraic manipulation is based on fundamental mathematical operations which are absolutely justified. The second method is justified by the fact that these transformations should be symmetric since choosing which frame is primed and which frame is unprimed is arbitrary and hence the primed and unprimed labels can be interchanged; however because the relative velocity is symmetric in magnitude but anti-symmetric in sign, its sign should be reversed while its magnitude should be preserved when exchanging the primed and unprimed labels.

2. Write down the Galilean transformations for space coordinates and time from O' frame to O frame where these frames are in a state of standard setting.
   Answer:
   $$\begin{aligned} x &= x' + vt' \\ y &= y' \\ z &= z' \\ t &= t' \end{aligned}$$

3. Two frames of reference, O and O', are in a state of standard setting where O' is moving along the common x-x' axis with velocity $v = 9$ relative to O (i.e. O' is moving in the positive x direction with speed 9). A ball which is at rest in frame O' is seen to be at position $(2.5, -3, 7.8)$ in this frame at time $t' = 0$. (a) Find the space coordinates and time of this ball in frame O at $t' = 10$ according to classical mechanics. (b) Repeat the question assuming that $v = -9$ (i.e. O' is moving in the negative x direction with speed 9).
   Answer:
   (a) Using the Galilean transformations of space and time from the primed to the unprimed variables, we have:
   $$\begin{aligned} x &= x' + vt' = 2.5 + (9 \times 10) = 92.5 \\ y &= y' = -3 \\ z &= z' = 7.8 \\ t &= t' = 10 \end{aligned}$$

   (b) Everything is the same as in (a) except x which will be:
   $$x = x' + vt' = 2.5 + (-9 \times 10) = -87.5$$

1.9.1 Space and Time Transformations

4. An inertial frame of reference O' is seen in another inertial frame of reference O to have its origin at $\mathbf{r} = (3.9, -12.3, 6.1)$ at $t = 0$, where at $t = 0$ we have $t' = 10$ with t and t' being those of O and O' respectively. Put these frames in a state of standard setting assuming that the corresponding coordinate axes in these frames are parallel and have the same orientation. Also assume that the other conditions for standard setting (e.g. uniform motion along the x-x' axis only) are satisfied.
   Answer: To put these frames in a state of standard setting we need to translate the origin of spatial coordinates of O' by a displacement vector $\mathbf{d} = (-3.9, 12.3, -6.1)$ and reduce the time t' by 10 so that at $t = 0$ the two origins will be coincident and the clock in O' reads $t' = 0$. Alternatively, we can translate the origin of O by a displacement vector $\mathbf{d'} = (3.9, -12.3, 6.1)$ and reduce the time t' by 10. The first adjustment represents bringing O' spatially to O while the second adjustment represents bringing O to O'. There are other more complex possibilities which the reader may wish to consider and investigate.

5. Two inertial frames, O and O', are in a state of standard setting where at $t = 3$ the origin of O' is seen by O to be at $x = 20$ and at $t' = 15$ the origin of O is seen by O' to be at $x' = -44$. Find the Galilean transformations of space coordinates and time between these frames.
   Answer: The times in the two frames are identical and hence at $t' = t = 15$ the origin of O is seen by O' at $x' = -44$. Moreover, if the origin of O is seen by O' to be at $x' = -44$ then the origin of O' is seen by O to be at $x = 44$. Hence, at $t = 3$ the origin of O' is seen by O to be at $x = 20$ and at $t = 15$ the origin of O' is seen by O to be at $x = 44$. Accordingly, we have:
   $$v = \frac{44 - 20}{15 - 3} = 2$$
   Therefore, we have:
   $$x' = x - 2t$$
   $$x = x' + 2t'$$
   while the transformations of the other space coordinates and time are as given before, that is:
   $$y = y'$$
   $$z = z'$$
   $$t = t'$$

6. A vehicle is traveling along a straight line with a constant velocity $v = 8$. An on-board siren sends a signal every 10 seconds. What is the distance traveled on the ground by the vehicle during that time interval as measured by an on-board observer and by an on-ground observer according to classical mechanics?
   Answer: First, although it is not stated, it is obvious that this problem can be modeled as occurring between two inertial frames (on-board and on-ground observers) which are

1.9.2 Velocity Transformations

in a state of standard setting where the uniform motion is taking place along the common x-x' axis. Second, since the question is "according to classical mechanics", then the appropriate transformations are the Galilean. Third, in such problems that involve difference between coordinates (since distance is the magnitude of difference between spatial coordinates) at two instants of time, the same result should apply to both frames since the positional separation as a result of movement in space and time involves the relative speed and time interval and both of these are common to the two frames according to the Galilean transformations of classical mechanics, that is:

$$d = d' = |v\Delta t| = |v\Delta t'| = |8 \times 10| = 80$$

where d and d' are the distance as measured by the on-ground observer and by the on-board observer.

7. O' and O are two inertial observers in a state of standard setting with $v = 45$. According to O, the distance between two events that occur at $t = 10$ is $d = 450$. What is the distance according to O' in classical mechanics?

   Answer: This question is similar to the previous question. However, while the previous question is about the positional separation between two events that occur at different instants of time, this question is about the positional separation between two events that occur at the same instant of time. So, it is more obvious that the two observers should agree on the spatial separation according to the Galilean transformations because on taking the difference between the coordinates, the vt term will drop out and hence the difference will be identical in both frames, that is:

$$d' = |x'_2 - x'_1| = |(x_2 - vt) - (x_1 - vt)| = |x_2 - x_1| = d = 450$$

where t (rather than t_2 and t_1) is used in both transformations because the two events occur at the same time. The numbers 45, 10 and 450, which are arbitrary, are chosen to confuse the reader and make the question more difficult. Like the previous question, the Galilean transformations are used because of the "classical mechanics" label.

1.9.2 Velocity Transformations

1. How are the Galilean transformations of velocity obtained from the Galilean transformations of space coordinates and time?

   Answer: By taking the first derivative of spatial coordinates with respect to time because velocity is the rate of change of position with respect to time. For example, for the transformations from O' frame to O frame we have:

$$u_x = \frac{dx}{dt} = \frac{dx}{dt'} = \frac{d}{dt'}(x' + vt') = \frac{dx'}{dt'} + \frac{d}{dt'}(vt') = u'_x + v$$

$$u_y = \frac{dy}{dt} = \frac{dy'}{dt'} = u'_y$$

$$u_z = \frac{dz}{dt} = \frac{dz'}{dt'} = u'_z$$

1.9.2 Velocity Transformations

where the Galilean time transformation $t = t'$, as well as the relevant spatial coordinate transformations, are used in these derivations.

2. Using the Galilean velocity transformations, show that if the momentum is conserved in an inertial frame then it is conserved in all inertial frames where you use a simple case of collision between two massive objects as a prototype in your demonstration. Repeat the question with the kinetic energy assuming this time a perfectly elastic collision.
Answer: Let have O and O' as two inertial observers in a state of standard setting with relative velocity v. Suppose that two massive objects, A and B, of mass M_A and M_B are seen by O to collide where they approach each other with velocities u_{A1} and u_{B1} and depart with velocities u_{A2} and u_{B2} where all these velocities are along the x axis. If the momentum is conserved according to O then we should have:

$$M_A u_{A1} + M_B u_{B1} = M_A u_{A2} + M_B u_{B2}$$

On applying the Galilean velocity transformations noting that mass is frame invariant, we obtain:

$$M_A \left(u'_{A1} + v\right) + M_B \left(u'_{B1} + v\right) = M_A \left(u'_{A2} + v\right) + M_B \left(u'_{B2} + v\right)$$

On canceling the v terms (i.e. $M_A v$ and $M_B v$) on both sides we obtain:

$$M_A u'_{A1} + M_B u'_{B1} = M_A u'_{A2} + M_B u'_{B2}$$

i.e. the momentum is also conserved in O' frame. Now, since O and O' frames are arbitrary, then this implies that the momentum is conserved in all inertial frames if it is conserved in any given inertial frame. In other words, the conservation of momentum is invariant under the Galilean transformations across all inertial frames.
Regarding the kinetic energy, which is also conserved since the collision is assumed to be perfectly elastic, we have:

$$\frac{1}{2} M_A (u_{A1})^2 + \frac{1}{2} M_B (u_{B1})^2 = \frac{1}{2} M_A (u_{A2})^2 + \frac{1}{2} M_B (u_{B2})^2$$

$$\frac{1}{2} M_A (u'_{A1} + v)^2 + \frac{1}{2} M_B (u'_{B1} + v)^2 = \frac{1}{2} M_A (u'_{A2} + v)^2 + \frac{1}{2} M_B (u'_{B2} + v)^2$$

Now, since the quadratic terms in v (i.e. $\frac{1}{2} M_A v^2$ and $\frac{1}{2} M_B v^2$) are common to both sides, they drop out. Also, because of the conservation of momentum, the mixed terms in u and v (i,e. $M_A u'_{A1} v$, $M_B u'_{B1} v$, $M_A u'_{A2} v$ and $M_B u'_{B2} v$) are equal on the two sides according to the first part of the question and hence they can also be canceled. Hence, we end up with:

$$\frac{1}{2} M_A (u'_{A1})^2 + \frac{1}{2} M_B (u'_{B1})^2 = \frac{1}{2} M_A (u'_{A2})^2 + \frac{1}{2} M_B (u'_{B2})^2$$

i.e. the kinetic energy is also conserved in O' frame. Again, since O and O' frames are arbitrary, then this implies that the kinetic energy in elastic collisions is conserved

1.9.2 Velocity Transformations

in all inertial frames if it is conserved in any given inertial frame. In other words, the conservation of kinetic energy in elastic collisions is Galilean invariant across all inertial frames.

We note that this example, both in its momentum part and in its kinetic energy part, is also an example of the principle of form-invariance of physical laws, i.e. invariance of conservation of momentum and invariance of conservation of kinetic energy in elastic collisions under the Galilean transformations.

We should also remark that the solution is meant to demonstrate (rather than prove) these invariance properties; otherwise several generalizations (e.g. when the masses or the number of objects before and after collision differ) are needed for a full proof although some of these generalizations are straightforward.

3. A head-on collision between two massive objects, A and B, is observed by two inertial observers O and O' who are in a state of standard setting with relative velocity $v = 1$ where the mass of the objects are $M_A = 2$ and $M_B = 3$. If according to O the initial velocities of A and B are $u_{A1} = 9$ and $u_{B1} = -2$ and the final velocity of A is $u_{A2} = 4$, what are the initial and final velocities of A and B according to O' assuming that all motions are taking place in one dimension?

    Answer: First, we need to know the final velocity of B according to O before we can fully answer the question. The final velocity of B according to O can be found from the conservation of momentum, that is:

$$
\begin{aligned}
M_A u_{A2} + M_B u_{B2} &= M_A u_{A1} + M_B u_{B1} \\
2 \times 4 + 3 \times u_{B2} &= 2 \times 9 + 3 \times (-2) \\
u_{B2} &= \frac{18 - 6 - 8}{3} \\
u_{B2} &= \frac{4}{3}
\end{aligned}
$$

    The initial and final velocities of A and B according to O' can be obtained from the Galilean velocity transformation, i.e. by just subtracting 1 from the O velocities, that is: $u'_{A1} = 8$, $u'_{B1} = -3$, $u'_{A2} = 3$ and $u'_{B2} = 1/3$. As we see, the momentum is conserved in both frames although the value of the momentum is different in the two frames where in O frame it is 12 (before and after collision) while in O' frame it is 7 (before and after collision). This is an example of form-invariance but not value-invariance since the momentum is conserved in both frames but its value is different in the two frames (refer to § 1.8).

4. A body A of mass $m_A = 5$ is seen by an inertial observer O to move with velocity $u_{A1} = 10$ along the x axis. Following an inelastic collision with an identical massive body B which is at rest in O frame, the two bodies coalesce and continue to move along the x axis. (a) What is the total momentum of this two-body system before and after collision according to O? (b) What is the velocity of the coalesced body after collision according to O? (c) What is the total momentum of the two-body system before and after collision according to another inertial observer O' who is in a state of standard setting with O where the relative velocity between O and O' is $v = 5$?

1.9.2 Velocity Transformations 31

Answer:
(a) Total momentum before collision in O frame:

$$m_A u_{A1} + m_B u_{B1} = 5 \times 10 + 5 \times 0 = 50$$

Total momentum after collision in O frame: the total momentum is conserved and hence the momentum after collision is equal to the momentum before collision, i.e. 50.
(b) Velocity of coalesced body after collision according to O: because the two objects coalesce and form a single object C with mass $m_C = m_A + m_B = 10$ (since mass is conserved in classical mechanics) that moves with a single velocity u_C then we have:

$$m_C u_C = 50 \quad \rightarrow \quad u_C = \frac{50}{m_C} = 5$$

(c) Total momentum before and after collision in O' frame: due to the conservation of momentum, the total momentum after collision is equal to the total momentum before collision. Now, after collision the object C is at rest with respect to O' because $u_C = v = 5$ and hence its momentum is zero. Therefore, the momentum of this two-body system according to O' is zero before and after collision. The reason that the total momentum is seen by O' to be zero even before collision (where A and B are seen by O' to be moving towards each other) is because in O' frame A is seen, according to the Galilean velocity transformation, to be moving with velocity:

$$u'_{A1} = u_{A1} - v = 10 - 5 = 5$$

while B is seen to be moving with velocity:

$$u'_{B1} = u_{B1} - v = 0 - 5 = -5$$

and since $m_A = m_B$ the total momentum according to O' is zero before collision despite the motion of A and B according to O'. In fact, according to O' the two objects approach each other with identical speed where they collide, bond and stop. We note that although the two observers agree on the conservation of momentum, their measured values of the momentum of the two-body system are different, i.e. 50 in O frame and 0 in O' frame. So, it is another example of form-invariance but not value-invariance. We also remark that since the collision is inelastic, the kinetic energy is not conserved in any frame although both agree on the amount of the energy lost, which is 125, due to the invariance of energy in the form of non-kinetic energy (e.g. heat) which is not frame dependent since it is independent of the relative motion.
5. Repeat the previous exercise with the kinetic energy (instead of momentum) assuming this time the collision is perfectly elastic.
Answer:
(a) Total kinetic energy before collision in O frame:

$$\frac{1}{2} m_A (u_{A1})^2 + \frac{1}{2} m_B (u_{B1})^2 = \frac{1}{2} \times 5 \times 10^2 + \frac{1}{2} \times 5 \times 0^2 = 250$$

Total kinetic energy after collision in O frame: the total kinetic energy is conserved in perfectly elastic collisions and hence the kinetic energy after collision is equal to the kinetic energy before collision, i.e. it is 250.

(b) Velocity of A and B after collision according to O: because the collision is elastic, the two objects do not coalesce. Now, from the conservation of kinetic energy in perfectly elastic collisions we have:

$$\frac{1}{2}m_A (u_{A2})^2 + \frac{1}{2}m_B (u_{B2})^2 = 250$$

and from the conservation of momentum we have:

$$m_A u_{A2} + m_B u_{B2} = m_A u_{A1} + m_B u_{B1} = 5 \times 10 + 5 \times 0 = 50$$

that is:

$$u_{A2} + u_{B2} = 10$$

since $m_A = m_B = 5$. On substituting from the last equation into the energy equation we obtain:

$$\begin{aligned}
\frac{1}{2}m_A (u_{A2})^2 + \frac{1}{2}m_B (u_{B2})^2 &= 250 \\
\frac{1}{2}m_A (u_{A2})^2 + \frac{1}{2}m_B (10 - u_{A2})^2 &= 250 \\
(u_{A2})^2 + (10 - u_{A2})^2 &= 100 \\
(u_{A2})^2 + 100 - 20u_{A2} + (u_{A2})^2 &= 100 \\
2(u_{A2})^2 - 20u_{A2} &= 0 \\
(u_{A2})^2 - 10u_{A2} &= 0
\end{aligned}$$

and hence either $u_{A2} = 0$ and $u_{B2} = 10$ or $u_{A2} = 10$ and $u_{B2} = 0$. Although both solutions are mathematically acceptable and they both conserve kinetic energy, the physically possible solution (where the two bodies are assumed to have a finite size, they collide and B is free to move) is $u_{A2} = 0$ and $u_{B2} = 10$. In fact, both solutions are physically acceptable if the two objects are indistinguishable.

(c) Total kinetic energy before collision in O' frame:

$$\frac{1}{2}m_A (u'_{A1})^2 + \frac{1}{2}m_B (u'_{B1})^2 = \frac{1}{2} \times 5 \times 5^2 + \frac{1}{2} \times 5 \times (-5)^2 = 125$$

Total kinetic energy after collision in O' frame: the total kinetic energy is conserved in perfectly elastic collisions and hence the kinetic energy after collision is equal to the kinetic energy before collision, i.e. it is 125. The velocities in O' frame are easily obtained from the velocities in O frame using the Galilean transformations where these objects are seen by O' to exchange their velocities, i.e. $u'_{A2} = -5$ and $u'_{B2} = 5$.

6. Using the Galilean velocity transformations, show that the momentum transforms between inertial frames by a constant additive difference.

1.9.3 Velocity Composition 33

Answer: The velocity of an object in frame O and frame O', where these frames are in a state of relative uniform motion with a constant velocity $\mathbf{v} = (v_x, v_y, v_z)$, is Galilean transformed as:

$$\mathbf{u} = \mathbf{u}' + \mathbf{v}$$
$$(u_x, u_y, u_z) = (u'_x, u'_y, u'_z) + (v_x, v_y, v_z)$$

On multiplying both sides of the last equation by the mass m, which is a frame invariant constant, we obtain:

$$(p_x, p_y, p_z) = (p'_x, p'_y, p'_z) + m(v_x, v_y, v_z)$$
$$\mathbf{p} = \mathbf{p}' + m\mathbf{v}$$
$$\mathbf{p} = \mathbf{p}' + \mathbf{C}$$

where $\mathbf{C} = m\mathbf{v}$ is a constant.

1.9.3 Velocity Composition

1. A predator is chasing a prey along a straight path where the velocity of the predator and prey are 13.5 and 12. What is the velocity of each relative to the other?
   Answer: Although it is not stated explicitly, it is obvious that these measurements are taken by an observer who is standing on the ground. Now, if we label the observer, the predator and the prey with "1", "2" and "3" and consider the motions to be along the x axis then we have:

   $$u_{x31} = u_{x32} + u_{x21}$$
   $$12 = u_{x32} + 13.5$$
   $$u_{x32} = 12 - 13.5$$
   $$u_{x32} = -1.5$$

   i.e. the velocity of the prey relative to the predator is $u_{x32} = -1.5$. Consequently, the velocity of the predator relative to the prey is $u_{x23} = -u_{x32} = 1.5$.

2. A radioactive nucleus ejects two beta particles in opposite directions where the speed of each one of these particles in the rest frame of the nucleus is $0.6c$. What is the speed of each one of these particles in the frame of the other particle according to classical mechanics?
   Answer: According to the Galilean transformations, the velocity composition is additive and hence the speed of each one of these particles in the frame of the other particle is $1.2c$. This can be obtained formally by using the above formula, as we will show next, but because the problem is very simple and the result is obvious, the answer can be obtained directly. If we label the two beta particles with "1" and "2" and label the nucleus with "3" and consider the motions to be along the x axis then we have:

   $$u_{x31} = u_{x32} + u_{x21}$$
   $$0.6c = -0.6c + u_{x21}$$
   $$u_{x21} = 1.2c$$

Hence, the speed of each particle relative to the other particle is: $|u_{x21}| = |u_{x12}| = 1.2c$.

3. Two inertial frames of reference, O and O', are in a state of standard setting with $v = 25$. An object is seen in O' to have a velocity component in the negative x direction of magnitude $u = 6.8$. What is the velocity of this object in O according to classical mechanics?
   Answer: If we label O, O' and the object with "1", "2" and "3", then according to the Galilean velocity composition formula we have:
   $$u_{x31} = u_{x32} + u_{x21} = -6.8 + 25 = 18.2$$
   The velocity components in the y and z directions are the same in both frames because the two frames are in a state of standard setting.

4. What are the y and z versions of the velocity composition formula according to classical mechanics?
   Answer: Following the x version of the velocity composition formula, we can similarly write:
   $$u_{y31} = u_{y32} + u_{y21}$$
   $$u_{z31} = u_{z32} + u_{z21}$$
   However, in the state of standard setting the relative velocity between the two frames (labeled as "1" and "2") in the y and z directions is zero and hence $u_{y21} = u_{z21} = 0$. Accordingly, the above formulae will reduce to:
   $$u_{y31} = u_{y32}$$
   $$u_{z31} = u_{z32}$$
   which are no more than another form of the formulae:
   $$u_y = u'_y$$
   $$u_z = u'_z$$

1.9.4 Acceleration and Force Transformations

1. What is the meaning and significance of the Galilean transformations of acceleration?
   Answer: The meaning is that all inertial observers agree on the acceleration of any given object, and hence as far as the inertial observers are concerned the acceleration is absolute. Now, since acceleration is an extrinsic property and hence it should be referred to a frame of reference, this should suggest the existence of a distinct unique frame (or absolute frame) that is common to all inertial observers. This unique frame, to which all accelerations are referred, unifies all the inertial observers in being so. Whether this frame is the frame of absolute space or the frame of ether or the frame that is established by the particular distribution of mass and energy in the Universe is unimportant. What is important is its presumed existence that unifies all the inertial observers.

1.9.4 Acceleration and Force Transformations 35

2. How are the Galilean transformations of acceleration obtained from the Galilean transformations of space coordinates and time?
   Answer: By taking the second derivative of space coordinates with respect to time because acceleration is the temporal rate of change of velocity which is the temporal rate of change of position. For example, for the transformations from O' frame to O frame we have:

   $$a_x = \frac{d^2x}{dt^2} = \frac{d^2x}{dt'^2} = \frac{d^2}{dt'^2}(x' + vt') = \frac{d^2x'}{dt'^2} = a'_x$$

   $$a_y = \frac{d^2y}{dt^2} = \frac{d^2y'}{dt'^2} = a'_y$$

   $$a_z = \frac{d^2z}{dt^2} = \frac{d^2z'}{dt'^2} = a'_z$$

   where the Galilean time transformation $t = t'$, as well as the relevant space coordinate transformations, are used in these derivations.

3. How are the Galilean transformations of force obtained?
   Answer: In classical mechanics, the mass is an intrinsic frame-independent constant and hence according to Newton's second law we have:

   $$\mathbf{f} = \frac{d\mathbf{p}}{dt} = \frac{d(m\mathbf{u})}{dt} = m\frac{d\mathbf{u}}{dt} = m\mathbf{a}$$

   where $\mathbf{f}$, $\mathbf{p}$, $\mathbf{u}$ and $\mathbf{a}$ are force, momentum, velocity and acceleration vectors while m and t are mass and time. This equation means that the force is proportional to the acceleration. Consequently, the Galilean transformations of force should follow the Galilean transformations of acceleration, i.e. the force is identical across all inertial frames that is:

   $$f'_x = f_x$$
   $$f'_y = f_y$$
   $$f'_z = f_z$$

4. An object of mass $m = 0.5$ is seen in an inertial frame O' at $t' = 1.3$ to be at rest in position $\mathbf{r}' = (3.5, 4.6, 1.3)$. At that time a force $f'_x = 6$ in the x direction is applied and the object is accelerated for 3 time units. Using classical mechanics, find the position, velocity, force and acceleration of the object at time $t = 10$ in another inertial frame O which is in a state of standard setting with frame O' where O' moves with velocity $v = 5$ relative to O. Also, find the applied force and the acceleration of the object in frame O during the application of force.
   Answer: We first solve the problem for frame O' and then we use the Galilean transformations to obtain the solution in frame O. According to Newton's second law, the acceleration during the application of force is given by:

   $$a'_x = \frac{f'_x}{m} = \frac{6}{0.5} = 12$$

1.9.4 Acceleration and Force Transformations

Accordingly, at $t' = 1.3 + 3 = 4.3$ we have:[7]

$$\begin{aligned}
x'(t' = 4.3) &= x'(t' = 1.3) + u'_x(t' = 1.3)\Delta t' + \frac{1}{2}a'_x(\Delta t')^2 \\
&= 3.5 + (0 \times 3) + (0.5 \times 12 \times 3^2) \\
&= 57.5 \\
u'_x(t' = 4.3) &= u'_x(t' = 1.3) + a'_x\Delta t' = 0 + (12 \times 3) = 36
\end{aligned}$$

where a'_x is the acceleration during the application of force.
At $t' = 10$ we have:

$$\begin{aligned}
x'(t' = 10) &= x'(t' = 4.3) + u'_x(t' = 4.3)\Delta t' + \frac{1}{2}a'_x(\Delta t')^2 \\
&= 57.5 + 36 \times (10 - 4.3) + 0.5 \times 0 \times (10 - 4.3)^2 = 262.7 \\
u'_x(t' = 10) &= u'_x(t' = 4.3) + a'_x\Delta t' = 36 + 0 \times (10 - 4.3) = 36 \\
f'_x(t' = 10) &= 0 \\
a'_x(t' = 10) &= 0
\end{aligned}$$

where a'_x is the acceleration after the application of force (i.e. not during the application) and hence it is zero.
Hence, at $t = t' = 10$ we have:

$$\begin{aligned}
x(t = 10) &= x' + vt' = 262.7 + (5 \times 10) = 312.7 \\
u_x(t = 10) &= u'_x + v = 36 + 5 = 41 \\
f_x(t = 10) &= f'_x = 0 \\
a_x(t = 10) &= a'_x = 0
\end{aligned}$$

As for the y and z components of position and velocity, they are identical in the two frames, that is at all times $t = t'$ we have:

$$\begin{aligned}
y &= y' = 4.6 \\
z &= z' = 1.3 \\
u_y &= u'_y = 0 \\
u_z &= u'_z = 0
\end{aligned}$$

Regarding the applied force and the acceleration of the object in frame O during the application of force, they are the same as in frame O' that is:

$$\begin{aligned}
f_x &= f'_x = 6 \\
a_x &= a'_x = 12
\end{aligned}$$

---

[7] We use t and t' for time (i.e. instant) and Δt and $\Delta t'$ for time interval. We also note that the above equations are based on the kinematics of uniformly accelerated motion which is fully explained in any standard textbook on general physics.

1.9.4 Acceleration and Force Transformations

5. A girl on a train throws a stone upwards with initial speed $u_0 = 3$. If the train is moving with velocity $v = 8$ along a straight railway, what is the position of the stone, as a function of time during the stone flight, as seen by the girl and as seen by a bystander on the platform who is opposite to the girl at the instant of throwing the stone?

   Answer: If we consider the frame of the platform as O and the frame of the train as O' and take the movement of the train to be in the x direction and upwards to be the y direction, then O and O' can be seen as two inertial frames in a state of standard setting with $v = 8$ where the event of throwing the stone occurred at the origin of coordinates at $t = t' = 0$. On using the equations of uniformly accelerated motion, the position of the stone as a function of time during the flight as seen by the girl can be found, that is:

$$\begin{aligned} x'(t') &= 0 \\ y'(t') &= u_0 t' + \frac{1}{2} a' t'^2 = 3t' - \frac{1}{2} g t'^2 \\ z'(t') &= 0 \end{aligned}$$

   where $g \simeq 9.81$ is the magnitude of the Earth gravitational acceleration.

   On applying the Galilean transformations, we obtain the position of the stone as a function of time during the flight as seen by the bystander, that is:

$$\begin{aligned} x(t) &= x' + vt' = 0 + 8t' = 8t \\ y(t) &= y' = 3t' - \frac{1}{2} g t'^2 = 3t - \frac{1}{2} g t^2 \\ z(t) &= z' = 0 \end{aligned}$$

6. Hooke's law for an ideal mass-spring system is given by:

$$m a_x = -k(x - x_0)$$

   where m is mass, a_x is acceleration, k is spring constant, x is mass position and x_0 is mass equilibrium position (for more details about Hooke's law, the reader is referred to general physics textbooks). Using the Galilean transformations, show that this law takes the same form in all inertial frames.

   Answer: Let have two arbitrary inertial frames, O and O', which are in a state of standard setting with relative velocity v. Assume that the above-given form of Hooke's law is valid in frame O. Now, on applying the Galilean transformations of mass, acceleration and x coordinate to the above form to transform it to frame O', we obtain:

$$\begin{aligned} m a_x &= -k(x - x_0) \\ m' a'_x &= -k([x' + vt'] - [x'_0 + vt']) \\ m' a'_x &= -k(x' - x'_0) \end{aligned}$$

   i.e. the law takes the same form in frame O'. Now, because O and O' are arbitrary inertial frames, then this applies to all inertial frames and hence Hooke's law takes the

1.9.4 Acceleration and Force Transformations

same form in all inertial frames, i.e. it is form invariant under the Galilean transformations. We note that k is an intrinsic property of the spring and hence it is frame independent, i.e. it is neither primed nor unprimed. This also applies to the mass in classical mechanics, but we used m' for more clarity.

7. Show that Newton's second law in its form: $f = ma$ is form invariant under the Galilean transformations.

   Answer: Let have two arbitrary inertial frames, O and O', which are in a state of standard setting with relative velocity v. Assume that the above form of Newton's second law (i.e. $f = ma$) is valid in frame O. Now, on applying the Galilean transformations of force, mass and acceleration to the above form of the law to transform it from frame O to frame O', we obtain:

$$\begin{aligned} f &= ma \\ f' &= m'a' \end{aligned}$$

   and hence this form of Newton's second law is form invariant under the Galilean transformations.[8]

8. Show that the more general form of Newton's second law: $f = \frac{dp}{dt}$ is also form invariant under the Galilean transformations.

   Answer: It was shown in the exercises of § 1.9.2 that according to the Galilean velocity transformations the momentum transforms between inertial frames by a constant additive difference, and hence we have:

$$p = p' + C$$

   where C is a constant. Therefore, under the Galilean transformations the above form of Newton's second law transforms between frames O and O' as follows:

$$\begin{aligned} f &= \frac{dp}{dt} \\ f' &= \frac{d(p' + C)}{dt'} \\ f' &= \frac{dp'}{dt'} + 0 \\ f' &= \frac{dp'}{dt'} \end{aligned}$$

   and hence this form of Newton's second law is also form invariant under the Galilean transformations.[9]

---

[8] We note that if the transformation of force is based on the transformation of acceleration (as we did in the text) then this sort of argument may be circular (or at least trivial), although it is useful for pedagogical purposes and for consistency check. However, this may be avoided by using other methods for establishing the Galilean transformation of force.

[9] Similar to the previous exercise, the validity (or usefulness) of this argument should depend on the method of establishing the Galilean transformation of force.

1.10 Newton's Laws of Motion

1. Name Newton's three laws of motion.
   Answer: They are: (a) the law of inertia, (b) the law of force-momentum, and (c) the law of action-reaction.
2. State Newton's first law descriptively and mathematically in a different form to the form given in the text.
   Answer: This law can be stated in many different forms. For example, it can be stated as follows: the vector sum of the external forces acting on a massive object is zero *iff* the acceleration of the object vanishes identically, that is:
   $$\mathbf{f}_s = \mathbf{0} \quad \Leftrightarrow \quad \mathbf{a} = \mathbf{0}$$
   where $\mathbf{f}_s$ stands for the vector sum of the external forces and $\mathbf{a}$ is the acceleration of the object.
3. State Newton's second law descriptively and mathematically.
   Answer: Newton's second law states that the force exerted on a massive object is proportional (or equal when proper units are chosen) to the time derivative of momentum, that is:
   $$\mathbf{f} = \frac{d\mathbf{p}}{dt}$$
   where $\mathbf{f}$ is the force exerted on the object, $\mathbf{p}$ is its momentum and t is time.
4. According to classical mechanics, what is the condition for writing Newton's second law in the form $\mathbf{f} = m\mathbf{a}$, i.e. force equals mass times acceleration?
   Answer: The condition is that m is constant, that is:
   $$\mathbf{f} = \frac{d\mathbf{p}}{dt} = \frac{d(m\mathbf{u})}{dt} = m\frac{d\mathbf{u}}{dt} = m\mathbf{a}$$
5. Discuss the issue of value-invariance of mass in classical mechanics and Lorentz mechanics.
   Answer: In classical mechanics the mass is value-invariant across all inertial frames, while in Lorentz mechanics it is value-invariant according to the modern convention and not value-invariant according to the old convention because it is a function of speed. However, all this is about the non-rest mass since the rest mass is value-invariant even according to the old convention of Lorentz mechanics.
6. Discuss the issue of the constancy of mass in classical mechanics and Lorentz mechanics with respect to speed.
   Answer: For a closed physical system (i.e. assuming there is no exchange of mass between the system and its environment), the mass is constant in classical mechanics. The mass is also constant (as a function of speed) in Lorentz mechanics according to the modern convention but it is a variable function of speed according to the old convention. Again, all this is about the non-rest mass since the rest mass is constant as a function of speed even according to the old convention.

7. What is the relation between Newton's first law and Newton's second law?
   Answer: Newton's first law is a special case of Newton's second law that corresponds to $\mathbf{f} = \mathbf{0}$ (or $\mathbf{a} = \mathbf{0}$). This can be seen from the formula of the second law:

   $$\mathbf{f} = \frac{d\mathbf{p}}{dt} = \frac{d(m\mathbf{u})}{dt} = m\frac{d\mathbf{u}}{dt} = m\mathbf{a}$$

   because when $\mathbf{f} = \mathbf{0}$, $\mathbf{u}$ should be constant and $\mathbf{a} = \mathbf{0}$ since $m \neq 0$ (i.e. uniform translational motion). Similarly, when $\mathbf{a} = \mathbf{0}$, $\mathbf{u}$ should be constant and $\mathbf{f} = \mathbf{0}$.

8. Is there any situation, according to classical mechanics, where the above form of Newton's second law (i.e. $\mathbf{f} = m\mathbf{a}$) does not apply?
   Answer: The form $\mathbf{f} = m\mathbf{a}$ does not apply when the mass is variable. According to classical mechanics, the mass of a closed system is constant (e.g. as a function of time or speed). However, we can imagine a change in the mass of a non-closed system by losing mass to outside, such as a rocket or a jet plane whose mass is diminishing due to the continuous ejection of gases and debris from its engine, or by gaining mass from outside such as a snow ball whose mass is increasing while rolling. In such cases the form $\mathbf{f} = \frac{d\mathbf{p}}{dt}$ must be used.

9. An object is seen to move along a straight line with a time dependent velocity given by $u = 3e^{-0.2t}$ while losing mass according to the relation $m = 6 - 0.01t$ where t is time. Find the force acting on the object at time $t = 5.3$ according to classical mechanics.
   Answer: This is a 1D motion which can be imagined to take place along the x axis. According to Newton's second law in its 1D form and the product rule of differentiation we have:

   $$\begin{aligned} f &= \frac{dp}{dt} \\ &= \frac{d(mu)}{dt} \\ &= u\frac{dm}{dt} + m\frac{du}{dt} \\ &= \left(3e^{-0.2t}\right)(-0.01) + (6 - 0.01t)\left(-0.6e^{-0.2t}\right) \\ &= -0.03e^{-0.2t} - 3.6e^{-0.2t} + 0.006te^{-0.2t} \\ &= (0.006t - 3.63)e^{-0.2t} \end{aligned}$$

   At $t = 5.3$ we have:

   $$f = (0.006 \times 5.3 - 3.63)e^{-0.2 \times 5.3} \simeq -1.247$$

   i.e. a force of magnitude $1.247\,\text{N}$ in the negative x direction.

10. State Newton's third law in simple words giving an example. What is the significance of this law?
    Answer: In simple words, Newton's third law is commonly stated as: for any action there is a reaction which is equal in magnitude and opposite in direction. As we see, "action" and "reaction" here are not technical terms. An example of this law is a person

1.10 Newton's Laws of Motion

exerting a force $\mathbf{f}_{pw}$ on a wall where the wall also exerts a force $\mathbf{f}_{wp}$ on the person where these forces satisfy the relation:
$$\mathbf{f}_{wp} = -\mathbf{f}_{pw}$$
The significance of this law is that in the physical world forces exist in opposite couples.[10]

11. Discuss the following quote: "The weakness of the principle of inertia lies in this, that it involves an argument in a circle: a mass moves without acceleration if it is sufficiently far from other bodies; we know that it is sufficiently far from other bodies only by the fact that it moves without acceleration".
    Answer: First, this seems to be based on the assumption that acceleration in space is felt due to the presence of mass in the neighborhood (which may be based on the Mach principle). But this is arguable and can be challenged. The situation will be different if we accept the existence of an absolute frame where acceleration is defined with respect to this frame. The principle of inertia will then be a special case of Newton's second law where force is defined as a physical agent that produces acceleration with respect to the absolute frame. Hence, we see no weakness in the principle of inertia if we accept the existence of an absolute frame.

12. Show by a simple qualitative non-rigorous classical argument that if Newton's laws are valid in a given frame of reference, then they should be valid in all frames of reference which are in a state of rest or uniform translational motion with respect to the given frame.
    Answer: All three Newton's laws of motion are about force and acceleration: the second law is about the relation between force and acceleration, the first is a special case of the second, and the third is about coupled forces. Considering that velocity is the time derivative of position and acceleration is the time derivative of velocity, we see that two frames which are in a state of relative rest or relative uniform motion should measure the same acceleration when they observe a given massive object because any constant difference between them in the position measurement will drop out by taking the first time derivative of position to obtain velocity, while any constant difference between them in the velocity measurement (due to their relative uniform motion) will drop out by taking the second time derivative of position to obtain acceleration and hence they should obtain the same acceleration measurement and consequently they should obtain the same force measurement (i.e. when they use a given physical formulation to correlate force to acceleration they will calculate the same value of force as a function of acceleration). Accordingly, if Newton's laws of motion (which are essentially laws about force and acceleration as stated above) hold true in one of these frames, these laws should also hold true in the other frames, as claimed.[11]

13. Show by a simple qualitative non-rigorous classical argument that Newton's three laws

---

[10] The third law (together with the second law) also implies the conservation of momentum.

[11] In fact, this argument may also require other assumptions about the objectivity and dependency of force. However, due to the pedagogical and non-rigorous nature of this argument, we do not go through these lengthy details. The reader can contemplate about these fundamental issues of physics and its philosophy.

of motion are form invariant under the Galilean transformations.
Answer: This question is very similar to its predecessor although it sounds more formal. It was shown in § 1.9.4 that according to the Galilean transformations, two inertial observers agree on their measurement of acceleration and force. Now, since Newton's three laws of motion are laws about forces and accelerations, these laws should be form invariant under the Galilean transformations. To be more clear, let assume that Newton's second law holds true in an inertial frame. Now, since all inertial frames agree on force and acceleration according to the Galilean transformations, then they should agree on Newton's second law, which is a particular relation between force and acceleration that involves mass, because mass is an intrinsic invariant property and hence it is the same for all observers. As for Newton's first law, it is a special case of Newton's second law and hence if the first law holds true in one inertial frame it should hold true in all inertial frames because Newton's second law holds true in all inertial frames. Regarding Newton's third law, it is about coupled forces and hence if it holds true in one inertial frame it should also hold true in all inertial frames because all these frames agree on forces according to the Galilean transformations. Consequently, all Newton's laws of motion will hold true in all inertial frames if they hold true in one inertial frame and hence they are form invariant under the Galilean transformations. The reader is also referred to § 1.9.4 for a more formal approach to this issue related to some instances (also see § 12.3.1).[12]

14. As stated in the text, Newton's laws of motion are supposed to be valid in Lorentz mechanics with some modifications in the definition of certain quantities and concepts. Now, someone may ask: what is the meaning of Newton's first law in the absence of absolute frame if we accept the view of special relativity that denies the existence of this frame?
Answer: It is difficult to have a legitimate meaning in the absence of absolute frame because what "rest" or "uniform motion", for example, will then mean? In fact, this in essence is the same as the challenge to special relativity about the justification of the definition of inertial and non-inertial frames and the difference between them where acceleration is difficult to define sensibly and realistically in the absence of absolute frame (see for example § 3.6.3 and § 9).

1.11 Thought Experiments

1. Discuss the main sources of error and traps in thought experiments.
   Answer: These include:
   • Treating these thought experiments, wittingly or unwittingly, like real world experiments and hence using them as evidence for purely hypothetical arguments which can be fundamentally wrong.
   • The thought experimenter can easily fall victim to his illusions, stereotypes and hid-

---

[12] We should remark that this argument is partly circular if we establish the Galilean transformation of force from Newton's second law, as we did in § 1.9.4. However, the purpose of this argument and its alike is pedagogical to highlight certain issues and hence the reader should not take it too seriously.

1.12 Requirements for Scientific Theories and Facts 43

den assumptions. In real world experiments, the physical reality will not allow this to happen although this danger still exists, but at a much lower level, in the preparations and in analyzing the results of real world experiments.

2. Give an example of thought experiments that are widely used in special relativity.
   Answer: Train thought experiment (see § 9.6.1).
3. Assess and criticize the use of thought experiments (or "thought methods" to be more general) in modern physics.
   Answer: In modern physics (and the relativity theories in particular) thought experiments are very common and may be seen as a clever way, or even fashion, for creating and establishing wonderful and robust physical ideas and arguments with very little cost or no cost at all. These thought experiments (or "thought methods" to be more general) may then be followed by searching for experimental and observational evidence in support of this thought physics, which may be called theoretical physics. Apart from being detached from reality and can lead to fatal illusions, this method of creating and developing physics makes the likelihood of twist, bias and even fabrication very high. In brief, if physics is to be about the real physical world then the theory should follow, rather than lead, the observation. At least, if it should lead, it should lead by a tiny margin where the observation should always be monitored during the creation and development of theory. A good example of this thought physics (whose roots can be traced to the popularity of the method of thought experiments) is the use of strings and membranes (which are utter fantasy with no experimental or observational evidence) to describe the physical world.

1.12 Requirements for Scientific Theories and Facts

1. Give examples of scientific theories and scientific facts.
   Answer: The answer to this question may not be straightforward as it might be thought because a theory in the view of a scientist may be an established fact in the view of another scientist. In fact, science is full of controversies and questionable issues and hence despite the commonly held belief that science is very solid and undisputed and it represents hard facts, there are many uncertainties not only in its interpretation but even in its formalism. A very brief inspection to the literature of scientific research will show that what is controversial in science is not less than what is noncontroversial. An example of scientific theory in the opinion of the author is general relativity. In fact, even the formalism of Lorentz mechanics (or at least part of it) regardless of any particular interpretation can also be considered as an example of scientific theory. Examples of scientific facts are the conservation of mass or energy or momentum or Newton's second law in classical mechanics (also see the answer to the next question).
2. In what "quantitative" sense scientific fact is a "fact"?
   Answer: This is another difficult and non-straightforward question. Apart from the controversies of science itself, the answer may also depend on the philosophical and epistemological views of the respondent. It is difficult to find a single fact in science that can be described as 100% fact (where 100% means exactly 100% with zero probability of

error). For example, the principle of conservation of mass can be verified within certain error margins.[13] So, in a given experiment we can say that mass is conserved within an error margin of 10^{-20} gram. This means that although the experimenter can guarantee that it is impossible to have a change of mass by more than 10^{-20} gram, he cannot rule out a possible change of 10^{-25} gram for example. This is because all experiments and experimental equipment have limited accuracy. This limited accuracy is essential both in classical physics and in modern physics due to the limitation on accuracy and resolution of equipment. In fact, the limits on accuracy and resolution have a more fundamental reason in some branches of modern physics (namely quantum mechanics) where the uncertainty in measurement is a fundamental principle of the observation process and is not a symptom of not having perfect tools and equipment. However, this issue may have a different interpretation from the one that we are interested in here. In fact, there are more fundamental sources of error in experiments than the limits on accuracy and procedural imperfection even in classical physics, that is the experiment can be wrong in its method, design, conduction and analysis and hence it could be fundamentally wrong. Considering the above example of mass conservation, we have implicitly assumed in our assessment that the experiment is approached, designed, conducted and analyzed correctly and hence it is 100% correct in these respects which is a difficult assumption considering the fact that there is always a possibility of error in the experiment itself and in its analysis and conclusions, i.e. it is wrong in principle. So, the simple conclusion is that nothing in science can be 100% correct in a mathematical sense although there are many things in science (as in daily life) that are 100% correct in a practical sense where the probability of being wrong is virtually (but not actually) zero. Yes, there may be some scientific issues of logical and mathematical nature that can enjoy this high level of mathematical certainty of being 100% true although this will be different from being true in a scientific sense by being obtained from experiment and observation of the physical world.

1.13 Speed

1.13.1 Speed of Projectile

1.13.2 Speed of Wave

1. Outline the main possibilities about the characteristic speed of projectile and the characteristic speed of wave.
   Answer: We have three main possibilities:
   (a) Projectile: the characteristic speed should belong to the rest frame of the source.
   (b) Wave in a medium: the characteristic speed should belong to the rest frame of the medium.

---

[13] We note that the principles of conservation of mass and conservation of energy (not mass-energy) still exist not only in classical mechanics but even in Lorentz mechanics where there is no conversion from mass to energy and vice versa (see § 4.4 and § 5.4).

1.13.2 Speed of Wave

(c) Wave without medium: there are several possibilities and hence the situation is not clear.

2. Outline the main possibilities about the observed speed of projectile and the observed speed of wave and their dependence on the relative motion of the observer, source and medium.
Answer: We have three main possibilities:
(a) Projectile: the observed speed depends on the relative motion between the source and the observer as well as the characteristic speed.
(b) Wave in a medium: the observed speed depends on the motion of the observer relative to the medium (but not on the motion of the source relative to the medium) as well as the characteristic speed.
(c) Wave without medium: there are several possibilities, e.g. the observed speed depends on the relative motion between the source and observer. There may also be the possibility that the speed is universal.

3. A buzzer is drifting in a waterway along a straight line where the water is running uniformly with velocity $\mathbf{v} = (8, 0, 0)$. There are two observers: A who is at position $\mathbf{r}_A = (-5, 0, 0)$ and B who is at position $\mathbf{r}_B = (10, 0, 0)$. At time $t = 0$ the buzzer, which is in the rest frame of the water, passes through the origin of coordinates. What is the observed speed of sound for A and B if the characteristic speed of sound in water is 1500?
Answer: We treat s_o, s_c and v as vectors and hence we have:
For A:
$$|s_o| = |s_c - v| = |-1500 - (-8)| = 1492$$
For B:
$$|s_o| = |s_c - v| = |1500 - (-8)| = 1508$$
We note that v is the velocity of the observer in the rest frame of the medium and hence we take it as -8 rather than $+8$.

4. Make an argument in support of the claim that the speed of a classical (or material) wave does not depend on the speed of its source.
Answer: Since the characteristic speed of a material wave is a property of the medium and depends on its physical characteristics, the wave becomes independent of the source as soon as it is emitted, i.e. the wave moves with its characteristic speed in the medium following its emission. This can be easily inferred from the fact that the speed of the sound source (e.g. plane jet) can reach and even exceed the speed of the emitted sound. For example, if the speed of the sound source is additive to the speed of the sound itself then breaking the sound barrier will not happen. Similarly, the sound of a fighter jet that moves at twice the speed of sound in air propagates at the speed of sound in air, i.e. the plane does not produce a sound that moves at twice the speed of sound in air or faster since the speed of sound is determined by the properties of the air (and possibly some properties of the sound itself like its frequency).

5. What can we conclude from the fact that the speed of sound inside a plane that moves at high speed is the same as the speed of sound in a room at rest.

Answer: The simple conclusion is that the characteristic speed of sound (or in fact any material wave) is relative to the rest frame of the medium of propagation.

1.14 Speed of Light

1. What is the difference between the characteristic speed of light and the observed speed of light?
   Answer: The characteristic speed of light is c which is a universal constant, while the observed speed of light is the actually measured value by individual observers and hence in principle it can be frame dependent and not necessarily constant across all frames. However, based on the analysis and interpretation of Lorentz mechanics it is commonly believed that the observed speed of light is always equal to the characteristic speed of light and hence it is a frame independent constant although the origin of this frame independence and the interpretation of the universality of the observed speed of light are contentious issues.

2. What are the main possibilities regarding the nature of light and its propagation according to classical physics? Also, how these possibilities affect the presumed speed of light?
   Answer: There are two main possibilities in classical physics about the nature of light that have a direct impact on its speed of propagation:
   (a) Light is a wave phenomenon and hence it propagates like a wave. Now, if we follow the analogy of material waves, like sound, which was generally presumed in classical physics, then it requires a medium through which it propagates. Accordingly, its characteristic speed will be relative to the rest frame of the medium (which is commonly known as ether) and hence the observed speed will depend on the movement of the observer relative to the medium.
   (b) Light is a particle (or corpuscular) phenomenon and hence it propagates like a projectile with no need for a medium of propagation. Accordingly, its characteristic speed will be relative to the rest frame of its source and hence its observed speed depends on the relative motion between the source and the observer. We note that according to this possibility the presence of a medium (like air or water) in the space of propagation may diminish the speed but this is not of interest to us in this context which is the speed of light in free space and hence we can assume that there is no medium or the presence of medium (like ether) does not affect the speed.
   We note that both these possibilities imply that the speed of light is frame dependent although the nature of this dependency is different between these possibilities, i.e. it depends on the movement of the observer frame relative to the medium frame in the first case while it depends on the movement of the observer frame relative to the source frame in the second case. We also note that the existence of absolute frame with no propagation medium like ether can be equivalent to the first possibility where the absolute space can play the role of propagation medium in providing the reference or rest frame for the characteristic speed although this (i.e. the existence of absolute frame with no propagation medium like ether) seems more compatible with the second possi-

1.14 Speed of Light

bility. Anyway, the above are just some possibilities among other possibilities that can be considered in analyzing the speed of propagation of light in free space.

3. Discuss and assess the validity of the Galilean transformations with regard to the two possibilities of the last question.

   Answer: The Galilean transformations are originally part of classical mechanics (as formulated in Newton's laws of motion) whose main focus is the movement of massive objects. Accordingly, if we assume that light is a corpuscular phenomenon it seems logical to assume that it follows the rules of classical mechanics and hence it should be subject to the Galilean transformations. However, if we consider light as a wave phenomenon then it may be less obvious and logical to assume that the Galilean transformations are the natural set of equations for light propagation. However, the success of classical mechanics at that time (i.e. before the emergence of Lorentz mechanics) made its rules and principles to be seen as very general and universal and hence all physical phenomena, including wave propagation phenomena, should follow these rules.[14] Consequently, the Galilean transformations were seen to apply to the wave phenomena (including light) as well as to the phenomena of projectile motion.

4. What is the other possibility (or possibilities) that emerged in modern physics about the propagation of light and its speed?

   Answer: The other main possibility (or possibilities) that emerged in modern physics is that regardless of the nature of light (i.e. being wave or particle) and regardless of any possible requirement of a propagation medium or not, the speed of light is independent of the movement of its source and its observer and hence it is frame independent. Although this may look like a unified view, it is not. Careful inspection of the literature of Lorentz mechanics in its early days (and even the literature of special relativity in the modern days) will reveal that there are significant differences about the origin and interpretation of this independence of frame and being a universal speed, e.g. some interpret this in its real sense while others interpret it in an apparent sense where this "apparent" may also have different interpretations. More details about these issues will follow in the upcoming parts of the book.

---

[14] This is amplified by the presumption of resemblance of material waves.

Chapter 2
Emergence of Lorentz Mechanics

2.1 Classical View of the World

1. Outline the main features and issues that characterize the classical view of the world as embedded in the philosophical and scientific framework of classical mechanics.
   Answer: The main features and issues of the classical view are:
   • The existence of a unique physical reality.
   • The existence of a unique truth which represents an accurate reflection of the unique physical reality.
   • The existence of space as a 3D manifold which contains all physical objects and events. This space is unique although it may be conceived in different forms by different observers and with respect to different physical settings.
   • The geometric nature of the space is Euclidean.
   • The space is universal so it is the same for all observers.
   • The space is absolute and hence it offers a frame of reference where states of rest and motion relative to this space can be sensibly and realistically defined.
   • The existence of time as a real, unique and evenly-flowing 1D continuum. Since time is 1D, it is necessarily Euclidean.
   • The time is universal and absolute so it is the same for all observers. Hence, all observers agree on the time of occurrence of any given event, and therefore they should agree on the order in time (i.e. before or simultaneous or after which may also be seen as past or present or future) of events and the duration (i.e. size of interval) of physical objects and series of events.
   • Space and time are independent of each other.
   • Space and time are independent of the physical objects and events. So, space and time can be considered as passive containers that embrace the physical objects and events. Also, space and time are independent of the state of rest or motion of the observer.
2. Search for some famous quotations about space and time in the Aristotelian philosophy and comment briefly.
   Answer: In this regard, we refer to some famous quotations attributed to Aristotle such as the following which is about time: "There is one single and invariable time, which flows in two movements in an identical and simultaneous manner ... Thus, in regard to movements which take place simultaneously, there is one and the same time, whether or not the movements are equal in rapidity ... The time is absolutely the same for both". Similar quotations about space can also be found in the literature.
   Comment: we note that this quote about time contains several elements of the classical view as described in the text, e.g. "one single and invariable time" which is an indication to its unique and absolute nature, or "which flows in two movements in an identical and simultaneous manner" which is an indication to its 1D evenly flowing nature and its

2.1 Classical View of the World 49

independence of the motion of observer.

3. Comment briefly on the following quote which is attributed to Newton about space: "Absolute space, in its own nature, without regard to anything external, remains always similar and immovable. Relative space is some movable dimension or measure of the absolute space, which our senses determine by its position to bodies, and which is vulgarly taken for immovable space ... Absolute and relative space, are the same in figure and magnitude; but they do not remain always numerically the same ... And so, instead of absolute places and motions, we use relative ones; and that without any inconvenience in common affairs ... etc.".

   Answer: The essence of this quote is that although there is an absolute space, it can only be measured and geometrized through the relative position of the objects inside this space and their particular distribution within it. So, "relative space" makes "absolute space" conceivable and measurable. We also note that this quote contains several elements of the classical view as described in the text, e.g. "without regard to anything external" which indicates its independence of the physical objects and events, or "remains always similar and immovable" which indicates its unique, passive and absolute nature.

4. Search for some famous quotations about time in classical mechanics and comment briefly.

   Answer: In this regard, we refer to the famous quotations of Newton such as the following: "Absolute, true, and mathematical time, of itself, and from its own nature flows equably without regard to anything external, and by another name is called duration ... Relative, apparent, and common time, is some sensible and external (whether accurate or unequable) measure of duration by the means of motion, which is commonly used instead of true time; such as an hour, a day, a month, a year".

   Comment: similar to what we saw in the quote of the previous question of "absolute space" and "relative space", we have "absolute time" and "relative time" where "relative time" makes "absolute time" conceivable and measurable. We also note that this quote contains several elements of the classical view as described in the text, e.g. "flows equably without regard to anything external" which indicates its evenly flowing nature and its independence of the physical objects and events.

5. Analyze the concept of universality of time in classical mechanics.

   Answer: The concept of universality of time in classical mechanics means that although different inertial observers may adopt different origins of time and different units of time scale, they can all be unified by simple translation to unify their origin of time and by simple scaling to unify their unit of time interval. Accordingly, the instant of time for the occurrence of any given event and the size of the time interval for the occurrence of a series of events or for the existence of an object will be the same for all inertial observers following this unification.

6. Discuss the claim that the classical concepts about space and time are intuitive and may even be instinctive and hence they may have biological roots in our mental blueprint.

   Answer: One thing in support of this view is that the development of philosophy (and natural philosophy in particular) as well as classical physics is based on these

2.1 Classical View of the World

classical concepts. All these early forms of knowledge have deep instinctive roots since they represent the first step in the generalization and organization of our daily life observations and experiences with some basic mental processes and logical additives all of which have obvious biological roots. Hence, the claim that the classical concepts of space and time are intuitive or instinctive with possible biological roots may not be far from reality as it might look.

7. Outline other features and issues in the classical view of the world which are not as general and recognized as the main ones but they have direct link to corresponding issues in modern physics and Lorentz mechanics in particular.
   Answer: Other issues include:
   • The basic physical properties of a physical object, such as mass and length, are independent of its state of rest or motion and hence they are frame-independent.
   • Mass and energy are two separate physical entities where each one is conserved on its own. Hence, the mass of a material object does not depend on its energy content (whether kinetic or non-kinetic).
   • Inertial frames and accelerating frames are defined with respect to an absolute frame which can be regarded as the frame of absolute space.
   • Light propagates in space like material objects and material waves. Accordingly, its observed speed is a variable that depends on the state of rest and motion (relative to the light source or relative to the propagation medium) of its observer, i.e. it is frame-dependent.

8. Name and discuss some of the other characteristics which are usually presumed (implicitly if not explicitly) about space in both classical and modern physics.
   Answer: Essential characteristics about the nature of space include homogeneity and isotropy where homogeneity means that the properties of the space are independent of the location (i.e. all positions of space have the same properties) while isotropy means that the properties of the space are independent of direction and orientation. We should remark that these characteristics refer to the properties of the space itself excluding any external factor and hence the space may lose its homogeneity or isotropy due to the presence of electric or magnetic fields, for example, but this does not affect the homogeneity and isotropy of the space itself because such inhomogeneity and anisotropy are caused and introduced by external agents (or rather belong to external agents). We should also exclude geometry-based gravitational theories, such as general relativity, which conceptualize and formulate gravity in terms of the curvature and geometric properties of spacetime and hence the homogeneity and isotropy of space will be lost in the presence of aggregates of matter and energy.

9. Name and discuss some of the implicitly or explicitly presumed characteristics of time in physics and science in general.
   Answer: Time in physics and science is generally understood to be a one-dimensional physical entity that is continuous, evenly flowing and pointing into a single direction (i.e. past $\to$ present $\to$ future but not the other way around or other different order).[15]

---

[15] Again, we should exclude geometry-based gravitational theories where some of the commonly-

We should note that these proclaimed characteristics of time may be considered to be based on the reality of time as a physical entity. In fact, the reality of time may be challenged and hence it may be seen as a synthesis of human brain. This, however, should not affect the concept of absolute time and its existence as a meaningful mental measure for the observed physical processes, i.e. the two are logically consistent.

10. Discuss the reasons for the success and failure of classical mechanics in relation to its reliance on the direct experience of mankind and daily practices.
    Answer: The success of classical mechanics can be partly explained by its reliance on common sense and daily experience and hence it represents generalizations of direct observations of mankind during his evolution. Accordingly, this mechanics was successful in describing and predicting the physical world and its diverse phenomena at scales and speeds that are commensurate to human perception of daily occurrences, i.e. sizes much larger than atomic size and speeds much lower than the speed of light. As a result, this mechanics has intrinsic limitations with respect to the size and speed of the observed phenomena and hence it could not cope with the challenges posed in dealing with physical phenomena at atomic scales and objects moving at very high speeds. These limitations led to the emergence of quantum mechanics to cope with the challenges of tiny size, and the emergence of Lorentz mechanics to cope with the challenges of high speed.

11. Name and discuss some of the principles of classical mechanics which are not in conflict with the framework of Lorentz mechanics (at least according to the common belief) but with the framework of quantum mechanics.
    Answer: The most prominent of these principles are determinism and preciseness where these principles reflect the highly realistic nature of classical mechanics. The essence of determinism is that a physical object has an independent reality which is determined in itself regardless of any observer, so the function of observation is to reveal this predetermined reality whereas in quantum mechanics the reality may be determined by the observation through the collapse of wave function. The essence of preciseness in classical mechanics is that there is no fundamental restriction or limit on the accuracy that can be achieved by measurements. Hence, according to the view of classical mechanics a measurement can be infinitely precise, at least *in principle*, so if we assume that we have ideal measuring equipment we can obtain exact measurements of physical quantities like momentum or energy. Based on this view, the error or uncertainty in our measurements is not because of an intrinsic physical requirement that puts restrictions on the accuracy of our measurements, but it is coincidental due to the limitation in our measuring equipment and procedures. However, according to the uncertainty principle of quantum mechanics the error or uncertainty is essential and intrinsic to the measurements of some physical quantities so *in principle* we cannot (rather than we can) have as a precise measurement as we wish even if we have ideal tools and measuring equipment. In fact, even the existence of such ideal tools may be or should be questionable if we have to follow the logic of quantum mechanics. So,

---

recognized properties of time may be dismissed or modified.

quantum mechanics puts inherent limits on both the reality (by denying determinism) and the truth (by denying preciseness) and hence it represents a fundamental departure from the very realistic view of classical physics which is based on these principles.

2.2 Galilean Relativity

1. State the Galilean relativity principle in a few and simple words. State any principal assumption about this principle. What is the obvious consequence of this principle?
   Answer: The Galilean relativity principle may be stated as: there is no privileged frame of reference where the laws of mechanics take a special form. The main assumption in this principle is that the frame is inertial since the laws of classical mechanics do not hold in non-inertial frames. The obvious consequence is that no experiment can be done to reveal the state of the frame as being at rest or in motion in absolute sense, i.e. relative to the presumed rest frame of absolute space. Hence, motion through absolute space can only be detected by acceleration and not by uniform velocity because space is indifferent to uniform velocity. Also, distinguishing mechanical effects depend on the acceleration of physical systems but not on their velocity, as we will see later.

2. Some authors may state the essence of the Galilean relativity principle as "absolute rest cannot be defined". Discuss this.
   Answer: We would rather say "absolute rest cannot be detected" while acceleration (i.e. with respect to absolute rest) can be detected. Hence, absolute rest can be defined although it cannot be detected (apart from the incomplete detection that is based on the detection of acceleration).

3. Analyze the Galilean relativity principle and its interpretation and the implication of this principle on the existence of absolute space.
   Answer: There are two main interpretations of the Galilean relativity principle, one is based on denying the existence of absolute space and the other on accepting this existence (although we believe only the latter should be accepted), that is:
   • There is no absolute space and hence the relativity principle applies because there is no sensible meaning (in an absolute sense) for being at rest or in motion. This interpretation requires justification for the physically-real difference between inertial and non-inertial frames where in the former the laws of mechanics hold true while in the latter they do not. In simple terms, being accelerating or non-accelerating (like being at rest or in uniform motion) is an extrinsic property and hence it requires a physically-real reference frame relative to which being accelerating or not can be attributed in a physically-real (rather than conventional) sense.
   • There is an absolute space and hence the essence of the relativity principle is that this absolute space does not distinguish in the applicability of the laws of mechanics between frames and if they are at rest or in motion as long as these frames are inertial, i.e. they are not accelerating relative to this absolute space. Yes, this space distinguishes between inertial frames and accelerating frames by denying the latter the privilege of holding the laws of mechanics. Accordingly, there is no logical inconsistency in this interpretation because it provides a physically-real reference frame relative to which being

2.2 Galilean Relativity

accelerating or not can be attributed and hence the distinction between inertial frames and accelerating frames, which is physically-real and observable, can be explained and justified by a real physical cause and will not be based on sheer convention.

4. Discuss the possibility that even in Lorentz mechanics the notion of absolute space can be accepted, i.e. there may be no contradiction between the formalism of Lorentz mechanics and the existence of absolute space in contrast to the view held by the special relativists that this mechanics put the final nail in the coffin of absolute space.
   Answer: Like classical mechanics, there is no contradiction between the formalism of Lorentz mechanics and the notion of absolute space and hence the second interpretation of the Galilean relativity principle, as discussed in the previous question, is consistent with the principles and formalism of Lorentz mechanics. Yes, the relativity principle of special relativity theory contradicts the existence of absolute space according to the general consensus of special relativists. However, special relativity is just an interpretation of the formalism of Lorentz mechanics and hence its principles and implications do not compel Lorentz mechanics to accept these principles and reconcile with these implications.

5. Is there a logical requirement for the theory of special relativity to deny the existence of absolute space, i.e. can we imagine special relativity to embrace the relativity principle in its classical sense?
   Answer: The reader is advised to consult the literature of special relativity. However, the general view is that special relativity is based on denying the existence of absolute space, either because it is a logical requirement or because it is a voluntary choice. As there are many contradictory views (and even flipflopping) in the literature of special relativity, we cannot provide a definite answer to this question. However, if special relativity adopted the relativity principle in its classical sense then the theory will lose one of its main ingredients and hence it may lose its status as special relativity.

6. What is the Newtonian principle of relativity? What is the Galilean invariance? What is the relation between them and the principle of Galilean relativity?
   Answer: The Newtonian principle of relativity and the Galilean invariance are just variant forms of the Galilean relativity principle or at least they are equivalent to this principle. The essence of the Newtonian principle and the Galilean invariance is that when the laws of motion of classical mechanics apply in a given reference frame then they equally apply in any other reference frame which is not in a state of acceleration relative to the given reference frame, i.e. knowing that a given reference frame is inertial, we can conclude that any reference frame which is in a state of rest or uniform motion relative to the given frame is also inertial.[16] So, by identifying a single inertial frame we can identify all other inertial frames by just observing their motion relative to the identified frame. In fact, by identifying a single inertial frame we cannot only identify all other inertial frames, but we can also identify all non-inertial frames as well by simple external observations. However, the opposite may not be true, i.e. by identifying a single accelerating frame we may not be able to identify the status of the

---

[16] We are following here the procedural definition of inertial frames.

other frames (at least easily by just observing their motion relative to the identified accelerating frame) and if they are inertial or not. The reason is that although we can identify the inertiality status of the other frame if it is in a state of rest or uniform motion relative to the identified accelerating frame since it should also be accelerating, we cannot easily identify its status by simple observations if it is accelerating with respect to the identified frame because two frames which are accelerating relative to each other can be both non-inertial or only one is non-inertial although they cannot be both inertial. For example, if one frame O_1 is moving (say relative to absolute space) with velocity $v_{x1} = 10$ and the other frame O_2 is moving with $v_{x2} = t$ then they are in a state of relative acceleration with the first being inertial while the second is not. But if we have a third frame O_3 that is moving with $v_{x3} = t^2$ then it is in a state of relative acceleration with both O_1 and O_2 (since the second time derivative of its velocity is not vanishing while the second time derivative of velocity of O_1 and O_2 are both vanishing) and it is obviously accelerating or non-inertial. Hence, relative to O_2 both O_1 and O_3 are accelerating but while O_1 is inertial O_3 is non-inertial. Accordingly, O_2 cannot determine by simple observations of the relative motion of O_1 and O_3 the status of these frames and if they are inertial or not. More details about this issue will be given in § 3.6.

7. What can you conclude from the essential difference between inertial and non-inertial frames?
Answer: The main conclusion is that distinguishing mechanical effects depend on the acceleration of the physical system and not on its velocity. So, the system can have any velocity but as long as the system is not accelerating it will hold the same laws of mechanics. However, its status will change dramatically, by violating the laws of mechanics, as soon as it experiences any form of acceleration, whether translational or rotational, uniform or non-uniform. So, acceleration is a fundamental distinguishing mechanical property of physical systems at least as far as classical mechanics is concerned.

8. Make a simple argument, based on the speed of light, for the case that according to classical physics which embraces the Galilean relativity principle, there is still an absolute frame (space or ether) and hence it is incorrect to claim that the relativity principle in its classical sense abolished absolute space but kept absolute time.
Answer: According to classical physics, the speed of light is subject to the normal addition rule of velocity composition as represented by the Galilean transformations (see § 1.9.2 and § 1.9.3). This should suggest that there is an absolute frame which is the frame in which the speed of light takes the special value c. So, the Galilean relativity principle should not imply the abolishment of absolute frame (space or ether).

9. Discuss the relation between the relativity principle, the Galilean transformations and Maxwell's equations.
Answer: As a consequence of the relativity principle, the Galilean transformations of space coordinates and time should be wrong if we accept that Maxwell's equations are the right set of equations for describing the electromagnetic phenomena and hence

these equations should be valid in all inertial frames by taking the same form.[17] This is because Maxwell's equations do not transform invariantly under the Galilean transformations and hence these transformations should be invalid. These issues will be discussed in detail in the following sections.

2.3 Maxwell's Equations and Speed of Light

1. Is it possible to challenge the claim that the equation $c = (\varepsilon_0 \mu_0)^{-1/2}$ does not refer to any particular frame of reference?
   Answer: Yes, this claim can be challenged because ε_0 and μ_0 are the absolute permittivity and permeability of "free space" and hence one may say that the frame of this space is the frame of reference that this equation refers to. So, if we assume that this free space is absolute where all relative motions are with respect to it, then c will be the speed of light in the frame of this absolute rest space. This, of course, depends on ascribing certain physical properties to this space some of which are the above permittivity and permeability as well as being the absolute rest frame where at least some of these properties cannot be denied by any one. In fact, if the ether hypothesis is accepted then ascribing these properties to the ether, which is supposed to be a fluid-like physical substance, is easier to imagine. Finally, we should remark that this answer does not address the issue of invariance under the Galilean transformations, which will be discussed in § 2.4, and hence even if this answer was accepted it cannot be a full solution to the problem (or problems) that emerged from Maxwell's equations.
2. Explain how the problem about the speed of light has emerged from Maxwell's equations. Also discuss how this problem was dealt with initially at that time.
   Answer: From Maxwell's equations the following equation for the speed of light c was obtained: $c = (\varepsilon_0 \mu_0)^{-1/2}$ where ε_0 and μ_0 (which are the permittivity and permeability of free space) are fundamental physical constants that can be measured from simple electric and magnetic experiments that do not involve measurements of speed or require a frame of reference. Now, since there is no explicit reference to a frame of reference in this equation, the following question then arises: to which frame of reference this speed of light belongs? This question, of course, is based on the classical view that the speed of light, like any other speed, is frame dependent.
   The problem was dealt with initially by assuming that the speed in the above equation is referred to the frame of ether whose existence, as a medium for the propagation of light waves, was generally recognized. This may have some basis even in Maxwell's equations themselves because ε_0 and μ_0 belong to the "free space" so the speed c belongs to this free space which can be easily identified with the frame of this "fluid-like" substance (i.e. luminiferous ether) that submerges the whole space.
3. Make a semi-formal classical argument that since Maxwell's equations contain the constant c, then Maxwell's equations are not form invariant under the Galilean transfor-

---

[17] We are assuming here that the relativity principle and the Galilean transformations should equally apply to all laws of physics and not only to mechanics. This extension apparently happened after Galileo during the rehearsal to the emergence of Lorentz mechanics.

mations.
Answer: Since, according to the Galilean transformations, speed in general, unlike acceleration, is frame dependent then the inclusion of c in the Maxwell's equations will make them form variant (instead of form invariant) under the Galilean transformations because c belongs to a particular frame and hence Maxwell's equations will not take the same form in another frame. However, this argument may be challenged because Maxwell's equations contain c which is a constant representing the characteristic (not the observed) speed of light and this constant is invariant across all frames. In fact, there are many details about this issue that are not worthwhile to investigate because the non-invariance of Maxwell's equations under the Galilean transformations can be established by a formal analytical argument, and hence this argument is redundant.

4. Referring to the previous question, as we still have the speed c incorporated in Maxwell's equations even after the replacement of the Galilean transformations with the Lorentz transformations, how the issue of form invariance of Maxwell's equations is then solved?
Answer: If we assume that the speed of light, which is represented in Maxwell's equations by c, is invariant (possibly in an apparent sense) under the Lorentz transformations and it does not belong to a particular frame then Maxwell's equations will take the same form because c will be the speed of light in any frame (not in a particular frame). In brief, the frame dependency of Maxwell's equations according to this challenge is based on the frame dependency of the speed of light, and hence if we make the speed of light frame independent then Maxwell's equations will be form invariant because it will contain a universal invariant speed rather than a frame dependent variant speed. We should also refer the reader to the potential challenge in the previous question about the status of c as the characteristic speed of light (which is just a universal constant) and not the observed speed of light and hence the invariance of Maxwell's equations should not be affected by the inclusion of c because c is the same for all frames even if the observed speed of light was frame dependent.

2.4 Maxwell's Equations and Galilean Transformations

1. Explain how another problem (this time related more directly to the Galilean transformations) has emerged from Maxwell's equations.
Answer: Because the laws of physics should be form invariant under the right set of transformations of space coordinates and time, Maxwell's equations were tested under the then accepted transformations of classical physics which are the Galilean transformations. These transformations proved to be the right ones with respect to the laws of classical mechanics since these laws transform invariantly under the Galilean transformations (refer to § 12.3). However, Maxwell's equations did not transform invariantly under the Galilean transformations. This led to question the validity of at least one set of these equations, i.e. either Maxwell's equations are wrong or the Galilean transformations are not the right transformations of space coordinates and time. The possibility of both sets (i.e. Maxwell's equations and Galilean transformations) being right was dismissed because of the demand of a unified physical theory, while the possibility of

both being wrong was dismissed due to the general acceptance of Maxwell's equations; moreover it is excessive since the problem can be fixed by assuming only one set is wrong.

2. Referring to the previous question, discuss in detail all the possibilities about the validity of the Galilean transformations and Maxwell's equations.

Answer: There are four main possibilities:

• Both the Galilean transformations and Maxwell's equations are correct. This was rejected because it is obviously illogical if we want a unified physical theory.

• Both the Galilean transformations and Maxwell's equations are wrong. Although this possibility is logical, it was also rejected for "economical reasons" because of the general tendency to reduce the amount of the effort required to fix a problem. So, as long as we have reasons to believe that both the Galilean transformations and Maxwell's equations contain some truth thanks to their experimental support, then let assume that only one of these is incorrect (or not exact) and hence focus on solving just one problem instead of two. Moreover, Maxwell's equations were well tested and enjoyed general acceptance.

• The Galilean transformations are right and Maxwell's equations are wrong. This possibility is logical and seemed to have some followers at that time. However, it was generally rejected due to the common belief in the validity of Maxwell's equations plus indications that the problem lies in the Galilean transformations.

• The Galilean transformations are wrong and Maxwell's equations are right. This possibility is logical and it is the one that eventually prevailed and led to the emergence and rise of Lorentz mechanics which is based on replacing the Galilean transformations with the Lorentz transformations which satisfy the form-invariance condition for Maxwell's equations as well as the laws of mechanics[18] (refer to the next question for more details; also see § 12.3). In fact, this possibility was endorsed by signs of other potential failures of the Galilean transformations apart from the problem of the invariance of Maxwell's equations.

3. Why the general opinion at that time seemed to accept Maxwell's equations and question the integrity of the Galilean transformations?

Answer: One reason is the strong validation and support to Maxwell's equations by the accumulation of massive experimental evidence. However, this reason may not be sufficient because the Galilean transformations have also strong support from the great success of classical mechanics. The magical power of Maxwell's equations and their elegance seemed to play part in this choice. In fact, Maxwell's equations were too sophisticated to get all that experimental endorsement and success by coincidence and sheer luck. On the other hand, classical mechanics is not characterized by such sophistication and preciseness and hence its success could be explained more easily by being a good approximation but not an exact and strictly correct formulation. In addition to all this, there were other signs and indications from other problems (e.g. the prob-

---

[18] In fact, the invariance of the laws of mechanics under the Lorentz transformations is made deliberately by introducing certain changes to the traditional form of classical mechanics to satisfy the condition of invariance.

2.5 Light as Wave Phenomenon and Luminiferous Ether 58

lem of Maxwell's equations and the speed of light) that the Galilean transformations may not be the right or the exact set of coordinate transformations. These suspicions were finally endorsed by the null result of the Michelson-Morley experiment (see § 2.6) which strengthened the possibility of the invariance of the speed of light and hence the invalidity of the Galilean transformations.

4. Discuss how the two problems that emerged from Maxwell's equations (i.e. the problem of having a specific speed of light without identification of frame of reference and the problem of Maxwell's equations being non-invariant under the Galilean transformations) are essentially identical.
Answer: This is because both these problems arise from the assumption of the validity of the Galilean transformations because the problem of having constant speed of light without identification of frame of reference is based on the assumption, which originates from the Galilean transformations and their additive nature regarding velocity composition, that this speed is frame dependent, while the problem of Maxwell's equations being non-invariant under the Galilean transformations is based on assuming that these transformations are the right set of transformations of space coordinates and time.

5. Discuss the issue that within the traditional domain of classical mechanics (i.e. low speed) the Galilean transformations of space coordinates and time are good approximation to the right (or exact) set of transformations which are the Lorentz transformations.
Answer: As we will see in § 4.5, at low speeds the Lorentz transformations converge to the Galilean transformations and hence the latter become good approximation to the former.

6. Discuss the possibility (which is similar to the first possibility in the text) that both Maxwell's equations and the Galilean transformations are right but Maxwell's equations apply to a privileged frame or ether frame.
Answer: This possibility was initially assumed but was eventually rejected due mainly to the null result of Michelson-Morley experiment (see § 2.6 and 12.2). In fact, this possibility also requires a fix for the problem of form invariance so that Maxwell's equations will be valid in all inertial frames. Also, this possibility is more about the problem of light speed which was investigated in § 2.3.

2.5 Light as Wave Phenomenon and Luminiferous Ether

1. Summarize the essence of this section about light as a wave phenomenon and the role of luminiferous ether outlining the relation of this to the concept of absolute space (or absolute frame).
Answer: As seen earlier, an issue related to the absolute space is the speed of light in vacuum and the existence of a medium in which the light propagates, the so-called luminiferous (i.e. light-bearing) ether. A common belief in the scientific community before the emergence of Lorentz mechanics was that the speed of light as a wave phenomenon is frame dependent and hence it should depend on the movement of the observer relative to the propagation medium (refer to § 1.13.2). The stated speed of

2.5 Light as Wave Phenomenon and Luminiferous Ether

light (i.e. $c = (\varepsilon_0 \mu_0)^{-1/2} \simeq 3.0 \times 10^8$ m/s where ε_0 and μ_0 are respectively the permittivity and permeability of free space) as obtained from Maxwell's equations (see § 2.3) then belongs to a privileged frame of reference that is the frame of the luminiferous medium. The space, which is at rest as observed from this privileged reference frame, may then be seen as the absolute space in which the speed of light is the one predicted by Maxwell's equations. Any reference frame O_2 which is in relative motion to this privileged reference frame O_1 should then measure a speed of light different from c, i.e. $c \pm v$ where v is the speed of O_2 relative to O_1.[19] Since the Earth moves around the Sun in an accelerated motion following an elliptical orbit, it was natural to try to measure its speed relative to the luminiferous medium which is supposed not to follow the Earth movement since it is unlikely that the Earth and ether frames are identical. This was the motivating background to which the Michelson-Morley experiment, among other experiments, was proposed and conducted to detect the ether wind and measure its speed relative to the Earth.

2. What are the two main types of wave? To which type of these light belongs?
   Answer: The two main types of wave are transverse waves and longitudinal waves. Light and other types of electromagnetic radiation are transverse waves.

3. Describe in general terms the role of ether in classical physics. Also, outline the essence of the ether theory and its logical implications on the speed of light.
   Answer: Because light propagates through space as waves, as can be inferred from Maxwell's equations and from showing signs of wave behavior like diffraction and interference, it was natural to assume that it, like other material waves such as water waves and sound waves, requires a medium through which it propagates. This hypothetical medium was called the luminiferous ether and it was believed to permeate the whole space.
   The essence and implications of the ether theory is that the propagation of light in space is like the propagation of sound through air or water where the characteristic speed of sound is with respect to the frame of medium (air or water) and hence its observed speed is subject to the Galilean rule of addition of velocities. So, we can imagine the ether according to the classical theory as a transparent medium (like thin air) that facilitates the propagation of light and all electromagnetic waves.

4. What is the meaning of "ether wind"?
   Answer: It means the movement of ether relative to the Earth where this movement is caused by the movement of the Earth through space which is the rest frame of ether, i.e. rotation of the Earth around its axis, rotation around the Sun, translation or rotation inside the Galaxy, movement as part of the Galaxy, etc.

5. Investigate the possibilities of the relation between ether and absolute space in classical physics.
   Answer: We do not have sufficient historical records about these issues; some of which may not have even been contemplated due to their pure hypothetical nature. However, we can list a number of possibilities about how the relation between ether and absolute

---

[19] Here, we are assuming a 1D motion along the line connecting the source to the observer.

2.5 Light as Wave Phenomenon and Luminiferous Ether

space may have been envisaged by classical physicists:
* Ether is at rest with respect to the absolute space. This seems to be the dominant view (if not the only view) at that time.
* Ether is in a state of relative uniform motion through absolute space.
* Ether is at rest or in a state of relative non-uniform motion which is dependent on the position in space and hence it could be at rest in one place and in motion in one direction in another place and in motion in another direction in a third place. If we add to this the possibility of temporal dependency of the ether motion, then the ether motion through absolute space will be time- and space-dependent.

We should remark that only the first view seems relevant and viable and hence the absolute space may be defined as the rest frame of the ether and hence the two frames are identical or equivalent.

6. Investigate the nature of ether and its role in the propagation of light according to the old electromagnetic theory.

   Answer: According to the available records, the ether in the old theory of electromagnetic propagation prior to the emergence of Lorentz mechanics possesses the following properties:[20]
   * It does not have the common properties of material media like mass and color.
   * It is completely transparent to light (and electromagnetic waves in general) and hence light transmitted through it will not suffer from dissipation. Also, the speed of propagation is independent of the frequency of light.
   * It permeates the whole space as well as all material objects (at least the transparent objects).
   * It does support transverse but not longitudinal waves.
   * It is perfectly elastic and extremely tenuous so that the planets and other physical objects do not lose energy by moving through it.

   However, there are a number of issues about the nature and physical properties of ether in the old theory which are not clear to us from the available records. For example, there is no conclusive record regarding the stand, prior to the emergence of Lorentz mechanics, about the type of medium through which light propagates in material media which are transparent to light like water or glass and if it should be the same or different type of luminiferous ether or the material medium itself and how this is related to the diminishing of the speed of light in such material media. It is also unclear if the luminiferous ether according to these views should also permeate the material media which are opaque to light and if so why light (or some part of the electromagnetic spectrum) cannot propagate through such media. In fact, it is difficult to find reliable historical accounts about these issues. Moreover, some of the available accounts are based on personal opinions and subjective judgments, while others may even be biased especially following the overwhelming rejection of the ether theory where this theory was ridiculed and treated as a typical example of pseudo-science or metaphysical beliefs.

---

[20] Some of these represent tentative claims that may be found in the literature, as will be indicated later.

2.6 Michelson-Morley Experiment

1. Outline the main proposals that were suggested to explain the null result of Michelson-Morley experiment.
   Answer: There were numerous proposals and explanations suggested by the scholars of that time to justify the null result of Michelson-Morley experiment, the main and most famous of these are:
   • Blaming the technical inadequacy of the experiment for the failure to detect the presumed effect and hence if this experiment was conducted properly the presumed effect should have been observed, as it is claimed by other scholars who conducted similar experiments which allegedly confirmed the ether wind proposal.
   • Blaming the reasoning and physical assumptions on which the experiment is based; more specifically the assumption that there is a relative movement between the Earth and the ether. According to the ether drag proposal, the Earth is supposed to drag the ether that surrounds the device and hence the part of the ether in which the light propagates in the Michelson-Morley experiment is in a stand still state with respect to the device and hence the experiment in principle cannot detect the ether wind because in the neighborhood of the apparatus there is no ether wind to detect. So, the whole logic and underlying assumptions of the Michelson-Morley experiment are flawed and that is why it failed.
   • FitzGerald-Lorentz length contraction which eventually led to the development of Lorentz transformations and Lorentz mechanics. The proposal in its original version is based on presuming a contraction of the apparatus (possibly due to presumed pressure from the ether wind)[21] in the direction of motion by an amount that exactly nullifies any difference in the time of the light journeys in the two arms of the device where this difference is supposed (according to the logic and analysis of the Michelson-Morley experiment) to be caused by the difference in the light speed due to the additive nature of velocity composition which is the working principle in classical physics even for the propagation of light. However, the interpretation of this length contraction in the original version and the role (and even the existence) of ether did change by the emergence of Lorentz transformations and Lorentz mechanics followed by the general acceptance and dominance of special relativity theory which disposed of ether and re-interpreted the nature and cause of this length contraction.[22]
2. What is the common factor that is shared between the problem of the speed of light that emerged from Maxwell's equations and the problem that emerged from the null result of the Michelson-Morley experiment and its analysis?
   Answer: The common factor is the presumption of the validity of the Galilean transfor-

---

[21] In fact, it is claimed in the literature that the ether exerts elastic stress on the moving objects.

[22] As we will see, length contraction alone cannot explain the null result of a more general type of Michelson-Morley experiment and hence time dilation is also needed to justify the null result. Accordingly, although the original length contraction proposal may be able to keep the frame-dependency of light (in principle at least and as a result of the Galilean transformations), this frame-dependency should eventually be discarded (with the Galilean transformations) if we add time dilation as well for a possible explanation for the null result.

2.6 Michelson-Morley Experiment

mations. This is because the speed of light, according to these transformations, should be frame dependent while Maxwell's equations apparently imply the universality of this speed (since the constant c which is inferred from Maxwell's equations is apparently not associated with any frame) and hence it is frame independent. Similarly, the null result of the Michelson-Morley experiment was a dilemma because the analysis of the result of this experiment was based on the assumption of the validity of the Galilean transformations, which required the dependency of the speed of light on the orientation of the apparatus, while the null result of this experiment seemed to indicate the independence of the speed of light from the orientation of the apparatus and hence the invalidity of the Galilean transformations.

3. What property should be attributed to ether if the ether drag hypothesis is to be sensible?

    Answer: We think the validity of the ether drag hypothesis requires attributing some sort of viscosity or/and resistance of penetration to material objects, i.e. the ether should stick to the equipment like viscous fluids and it should have limited penetration ability to material objects. Some of these properties may not be entirely compliant with the presupposed properties of ether.

4. Assess and criticize the Michelson-Morley experiment and similar experiments.

    Answer: It seems to the author that these experiments have been given more weight and credibility than they deserve. There were counter claims of similar experiments that confirmed the predicted effect or at least they did not reach the same firmly-confirmed null conclusion of the Michelson-Morley experiment and its alike.[23] Although we do not wish to suggest in this context that the Michelson-Morley experiment was flawed or it was based on wrong assumptions and analysis, we think more improved versions should have been conducted before reaching such a firm conclusion. Although there were subsequent improved versions of this experiment as well as other experiments, the majority of these experiments have been conducted with the motive to confirm the null result of Michelson-Morley experiment and were under the banner of special relativity, and hence it is very likely that they are affected by some sort of bias (at least non-intentional bias). Moreover, such a huge conclusion that shakes the foundations of physics should have not been reached from essentially a single experimental method that could be fundamentally wrong for some reason that no one is aware of. We think various methods that are totally different from the Michelson-Morley method and its physical principles and logic should have been invented and tried before declaring the null result as a solid fact. The main blame, however, should not be put on the scientists of that era who did their best by using the best available theoretical and experimental resources and techniques at that time to reach a conclusion about this fundamental issue, but it should be put on the modern scientists who should not have put such a massive weight and credit behind these historical experiments and their null conclusion. Accordingly, we think the contemporary scientists, who spend huge resources on petty experiments with trivial implications and conclusions, are invited to revisit this issue with a new approach

---

[23] We should also add the possible explanation of the claimed detection of a small phase shift in the Michelson-Morley experiment, as indicated earlier.

that is based on total objectivity and freedom from any influence of special relativity or any other theory. There are several claims from respected scientists about the variation and frame-dependency of the speed of light in astronomical observations and scientific experiments. These claims should be considered more seriously and hence they should not be dismissed without inspection or notice because of their conflict with special relativity and its postulates or because they are in conflict with the result of Michelson-Morley experiment.[24] Although we will reach a conclusion that the formalism of Lorentz mechanics suggests the invariance of the speed of light, new evidence based on a new approach to this problem may lead to amendments to the current formalism of Lorentz mechanics or at least could shed more light on this vital issue and many other related issues which are at the heart of modern physics.

Finally, we should remark that the Michelson-Morley experiment (as well as its alike) is restricted in its conclusion about the invalidity of the Galilean transformations because this experiment is based on a classical wave propagation model and hence it does not rule out the possibility of a classical projectile (or ballistic) propagation model or potentially other propagation models (refer to the exercises of § 12.2). This should endorse our view that these experiments have been given more weight and credibility than they deserve.

2.7 FitzGerald-Lorentz Proposal and Emergence of Lorentz Transformations

1. In appendix § 12.2 about the Michelson-Morley experiment, it was shown that the times for the light in its PP_1P journey and in its PP_2P journey are different according to the logic and analysis of Michelson and Morley. Using the derived expressions for these times and assuming $L_1 = L_2$, derive the FitzGerald-Lorentz length contraction formula. Comment on the result and how this length contraction effect is supposed to nullify the effect of any time difference in this analysis.

   Answer: In appendix § 12.2, it was shown that:

   $$t_1 = \frac{2cL_1}{c^2 - v^2} = \frac{2L_1}{c(1 - v^2/c^2)}$$

   $$t_2 = \frac{2L_2}{\sqrt{c^2 - v^2}} = \frac{2L_2}{c\sqrt{1 - v^2/c^2}}$$

   Now, if $L_1 = L_2$ we get:

   $$\frac{t_1}{t_2} = \frac{1}{\sqrt{1 - v^2/c^2}} = \gamma$$

---

[24] We should also refer in this context to a number of superluminal phenomena, mainly associated with astronomical observations, which are allegedly explained and justified within the framework of special relativity without violation of the invariance of the speed of light. We note that superluminal speed means speed exceeding the characteristic speed of light c.

This expression can be similarly derived from the times of the light journey in the two arms of the apparatus after rotation by 90°, that is:

$$\frac{T_2}{T_1} = \frac{1}{\sqrt{1 - v^2/c^2}} = \gamma$$

Now, if we have to nullify the effect of this time difference by a length difference (or contraction) then we should have:

$$\frac{t_1}{t_2} = \frac{l_2}{l_1} = \gamma \quad \Rightarrow \quad l_1 = \frac{l_2}{\gamma}$$

where l_1 and l_2 are the effective lengths (i.e. actual lengths which may be affected by the movement through ether) of the apparatus arms which are parallel and perpendicular to the ether wind. This is equivalent to having the length of the arm which is parallel to the ether wind (i.e. PP_1 whose apparent or proper length is L_1) being contracted by the γ factor. Similarly for the apparatus after rotation by 90° we have:

$$\frac{T_2}{T_1} = \frac{l_1}{l_2} = \gamma \quad \Rightarrow \quad l_2 = \frac{l_1}{\gamma}$$

which, again, is equivalent to having the length of the arm which is parallel to the ether wind (this time it is PP_2 whose apparent or proper length is L_2) being contracted by the γ factor. Now, since we are assuming $L_1 = L_2$, then we can change the labeling and write:

$$l = \gamma l'$$

where l and l' are the proper and improper length of the arm parallel to the ether wind (whether PP_1 or PP_2). The last formula is just the well known formula of length contraction of Lorentz mechanics. So, according to the FitzGerald and Lorentz analysis this length contraction effect which is suffered by the arm that is parallel to the ether wind (whether PP_1 or PP_2) will compensate by the exact amount to nullify the effect of the time difference that is derived from the Michelson-Morley experiment and analysis. We note that if we follow the FitzGerald and Lorentz analysis then there is no necessity for the denial of the existence of ether. According to the available historical records, Lorentz (at least) did not deny the ether in his theory. In fact, the FitzGerald-Lorentz proposal of length contraction was apparently suggested to salvage the ether not to deny it or abolish it. We should also remark that a sort of "ether pressure" on the moving object seems to have been envisaged as the cause for the proposed length contraction.

2. Outline in a few words the essence of the FitzGerald-Lorentz proposal of length contraction as can be understood from the analysis of the previous question.
Answer: Because the time of the light journey in the parallel arm is longer than the time of the light journey in the perpendicular arm by the γ factor (assuming that the two arms are of equal length), then to nullify the effect of this time difference we need to assume that the length of the parallel arm is contracted by the γ factor to make the time of the two journeys equal.

3. Investigate the problems that have been addressed by the proposal of Lorentz transformations.
   Answer: There are several problems and question marks that found a convincing solution (at least so far) by the proposal of Lorentz spacetime coordinate transformations, although not all these problems are fundamentally different and hence they can be considered as different faces of a single problem. Examples of these problems include explaining the null result of Michelson-Morley experiment and unifying the laws of mechanics (with some modifications) and the laws of electromagnetism in being form invariant under the proposed spacetime coordinate transformations. In fact, Lorentz himself in one of his papers showed that Maxwell's equations are invariant under the new set of spacetime coordinate transformations.
4. Make a brief comparison between the Galilean transformations and the Lorentz transformations (refer to § 1.9 and § 4.2.1).
   Answer: There are many things to be said in answering this question; some of these are listed in the following points:
   • Space and time are more mixed in the Lorentz transformations than in the Galilean transformations.
   • Time, as well as space, looks in the new set of transformations frame-dependent while in the Galilean transformations only space (or rather its coordinates) looks frame dependent.
   • A new factor (i.e. Lorentz factor γ) has emerged in the new set of transformations. This factor does not exist in the old Galilean transformations.
   • While only the laws of classical mechanics are form invariant under the Galilean transformations, both these laws and the laws of electromagnetism (represented by Maxwell's equations and other derived results like the electromagnetic wave equation) have this property of being form invariant under the Lorentz transformations (refer to § 12.3 for more details).
5. Can length contraction alone explain the null result of the Michelson-Morley experiment and its alike?
   Answer: It was shown in the literature that even if we assume length contraction in the direction of motion, the Michelson-Morley device, but with unequal arms, should still display a phase shift in the pattern (according to the Michelson-Morley method of analysis) over a period of six months as the Earth reverses its orbital direction around the Sun. In fact, the Kennedy-Thorndike experiment, which was conducted in the 1930s, was designed to investigate this possibility where the null result of this experiment concluded that length contraction alone cannot fully explain the failure of the Michelson-Morley experiment and hence time dilation (i.e. the full consequences of the Lorentz spacetime coordinate transformations in its spatial and temporal effects) as well is needed to explain the null result of the Michelson-Morley experiment and its alike (e.g. Kennedy-Thorndike experiment). However, we still need other assumptions, such as the assumption of an ether wind with a component in the plane of the device (see § 12.2), to reject the ether hypothesis and hence these experiments cannot rule out the possibility of an ether drag for instance. Also, these experiments cannot rule

out interpretations (like Lorenz interpretation) that are based on the existence of ether with the admission of the Lorentz spacetime coordinate transformations. So in brief, if the Kennedy-Thorndike experiment was correct and conclusive then it only implies that length contraction alone cannot fully explain the null result of a Michelson-Morley type experiment in more general settings although it may be sufficient to explain the null result of the original Michelson-Morley experiment where the arms of the device are equal within the allowed error margins.[25]

2.8 Poincare Suggestion and Subsequent Developments

1. Summarize the main historical events and scientific developments that led to the emergence and rise of Lorentz mechanics.
   Answer: We can summarize these events and developments in the following points:
   • According to the classical view of the world, time and space are absolute.
   • According to the Galilean relativity, which is part of the classical view, although there is an absolute space and hence an absolute rest frame, this frame cannot be identified by an experiment conducted in a reference frame which is at rest or uniform motion with respect to this rest frame. So, absolute space does not distinguish between inertial frames and whether they are at rest or in uniform motion with respect to the absolute frame although this absolute space demonstrates itself by discriminating against accelerating frames which are denied the privilege of holding the laws of classical mechanics.
   • The emergence of Maxwell's equations produced two problems. The first problem is that the constant c, which represents the characteristic speed of light in free space, emerged from Maxwell's equations as a constant that can be calculated from other two physical constants, ε_0 and μ_0, with no reference to any frame of reference. Now, since the classical view is based on assuming that the speed of light is frame dependent, a question has then emerged that to which frame the constant c belongs? The natural answer was that the constant c belongs to the rest frame of absolute space or to the rest frame of the "luminiferous ether" which is supposed to be a medium for the transmission of light that invades all space.
   • The second problem that emerged from Maxwell's equations is that although the laws of mechanics are invariant under the Galilean transformations of space coordinates and time (which are the legitimate transformations in classical mechanics) as they should be, Maxwell's equations failed to pass this test since they were not invariant under the Galilean transformations. Now, since the requirement of form invariance is essential for any objective and general physical theory, a question has been raised that which of the two sets (i.e. Galilean transformations and Maxwell's equations) is wrong?
   • Based on the Maxwell light speed problem and the presumed solution of the existence of ether, a natural quest for measuring the speed of the Earth in its motion relative to the ether then emerged. The prominent example of this quest is the Michelson-Morley

---

[25] We should also note that all these experiments cannot rule out a projectile propagation model that is consistent with the classical transformations since the analysis of the Michelson-Morley experiment and its alike is based on a wave propagation model.

2.8 Poincare Suggestion and Subsequent Developments

experiment to measure the speed of the "ether wind" which is presumably caused by the Earth motion through space using interferometric techniques. This experiment failed to detect the supposed effect from the presumed ether wind and hence the existence of ether was in question. The experiment also highlighted a number of problematic issues in classical physics such as the frame dependency of the speed of light and potential invalidity of the Galilean transformations.

• A proposal from FitzGerald and Lorentz of a length contraction effect in the direction of motion relative to absolute space or ether then appeared. The purpose of this proposal is to offset the effect of any difference in the time of the light journeys in the Michelson-Morley apparatus (see § 12.2 for details) through the compensation with length contraction by exactly the same amount required to annihilate the time difference and hence nullify the supposed phases shift of the interference pattern generated from the two interfering light beams in this experiment which is caused by this difference in the time of the light journeys. As we will see next, the length contraction proposal was developed further by Lorentz who introduced the Lorentz spacetime coordinate transformations which formed the basis for the theory and formalism of Lorentz mechanics.

• The second problem that emerged from Maxwell's equations, which was not addressed by the FitzGerald and Lorentz proposal of length contraction, was then addressed fundamentally by Lorentz who suggested a set of equations (to be known later as the Lorentz transformations) to transform space coordinates and time between inertial frames. These transformations proved to be the correct set of equations for invariantly transforming both the laws of classical mechanics (with some modifications) and Maxwell's equations. This suggestion represents the birth of Lorentz mechanics and its basic formalism which were further developed in the subsequent years and decades by many scientists to be the mechanics of Lorentz transformations in its modern form.

• Although the proposal of FitzGerald and Lorentz which seemed to solve the problem (partially at least) of failing to detect the ether by the Michelson-Morley experiment and its alike is not based on denying the existence of ether,[26] this existence was questioned by Poincare (and later by other scientists) who proposed a collection of philosophical and physical ideas and procedures which formed the framework and components of what will be known later as the special theory of relativity. Merging the ideas of Poincare (in the form of the elements and framework of special relativity) and the formalism of Lorentz (in the form of Lorentz transformations) with some added amplifications and extensions, Einstein came on the scene with his 1905 paper. This attempt by Einstein was not seen at that time as an exceptional among the collective effort in developing the theory of Lorentz mechanics, and hence the development of Lorentz mechanics (e.g. by scientists and mathematicians like Planck, Laue, Minkowski, etc.) continued under a much larger umbrella and a more flexible framework that is neutral to any particular interpretation. However, the situation changed suddenly and dramatically with the sudden rise of Einstein to fame following the alleged confirmation of the general theory

---

[26] The non-denial of ether also applies to the suggestion of Lorentz transformations by Lorentz.

of relativity by the 1919 solar eclipse expedition of Eddington and his team. From that point in history, Lorentz mechanics took a different route of development where it is seen as the mechanics of special relativity theory and hence all other interpretations of Lorentz mechanics were abandoned and even suppressed and all efforts to investigate possibilities that were excluded by special relativity theory were ridiculed.[27] Moreover, the whole credit for the emergence and development of Lorentz mechanics (which is a collective effort of many scientists before and after Einstein) was totally given to Einstein in a clear violation of all the historical facts and the ethical code of science. In fact, the role of several of these scientists (notably Lorentz and Poincare before and Minkowski after) is more important and pivotal than the role of Einstein.

2. Discuss the claim by some special relativists that even Lorentz attributed the credit for Lorentz mechanics (under the label of special relativity) to Einstein in the quote that is attributed to Lorenz whose essence is that: the relativity theory belongs to Einstein.
Answer: This is not an attribution of credit but it is a denial by Lorentz of his acceptance to the special relativity interpretation of Lorentz mechanics. So, what Lorentz was essentially saying is that: I do not believe in the special relativity interpretation of Einstein because I have my own interpretation which is different from the special relativity interpretation.

---

[27] We mean in the mainstream physics.

Chapter 3
Introduction to Lorentz Mechanics

3.1 Lorentz Mechanics versus Other Mechanics

1. Discuss the relation between classical mechanics on one hand and Lorentz mechanics and quantum mechanics on the other hand.
   Answer: In simple words, Lorentz mechanics was developed to rectify the defects of classical mechanics with regard to high speed, while quantum mechanics was developed to rectify the defects of classical mechanics with regard to small size. So, both Lorentz mechanics and quantum mechanics address limitations of classical mechanics related to speed and size respectively.
2. Discuss the difference between the relation of Lorentz mechanics to classical mechanics and the relation of Lorentz mechanics to quantum mechanics.
   Answer: As seen in the answer to the previous question, Lorentz mechanics was developed to address the limitations of classical mechanics with regard to high speed,[28] while quantum mechanics was developed to address the limitations of classical mechanics with regard to small size. Accordingly, Lorentz mechanics was originally developed to address certain issues in classical mechanics, but Lorentz mechanics was not originally developed to address certain issues in quantum mechanics. Consequently, the relation between Lorentz mechanics and classical mechanics is the relation between a general (or extended) theory and a special (or restricted) theory, while the relation between Lorentz mechanics and quantum mechanics is the relation between two different theories that have common elements where these two theories meet in investigating small objects moving at high speeds.
3. Discuss the relation between the special theory of relativity (as a representative of Lorentz mechanics) and the general theory of relativity.
   Answer: This issue is controversial where some scholars claim that special relativity is a special or limiting case for general relativity while others claim that the two theories are totally different and the only common feature between them is the "relativity" label which could be somewhat arbitrary. Our belief is that the latter opinion is more substantiated because of the following reasons:
   • The subjects of the two theories are very different. While special relativity is a theory of inertial frames,[29] general relativity is a theory of gravitation and hence the distinc-

---

[28] This, in a sense, should also include the limitation of classical physics to deal with electromagnetism.
[29] We could say more accurately that Lorentz mechanics, as represented by special relativity, is essentially a theory of the mechanics of motion as seen from inertial frames where the transformations of space coordinates and time are the focus of the investigation since these transformations underlie the whole physics of motion in its wide sense and mechanics in general and even to the more extended subject of physics which includes for example electromagnetism due to the fact that the concepts of time and space and their transformations are fundamental ideas with far reaching consequences in all branches of physics.

tion between the two is obvious. Inertial frames cannot be a special case of gravity. Such a claim may be based on the equivalence principle where acceleration and gravitation are alleged to be equivalent. However, this principle may be contested; moreover there are restrictions on its validity and application. In fact, this principle is more of an assumption (or axiom or postulate) than an established fact.
- The two theories have different sets of postulates, axioms and principles.
- The formalism of the two theories is completely different. No one has shown that the formalism of special relativity (in the form of Lorentz spacetime coordinate transformations and their derived consequences) can be obtained from the formalism of general relativity (in the form of field equations) as a special or limiting case as it is the case with classical mechanics in its relation to Lorentz mechanics where the formalism of the latter converges to the formalism of the former at low speeds (see § 4.5).
- The formulation of the two theories in terms of the geometry of spacetime manifold where this manifold is considered flat in special relativity and curved in general relativity (and hence the spacetime of special relativity is a special case to the spacetime of general relativity or the spacetime of general relativity is locally flat) is not sufficient to make special relativity a special case or an approximation to general relativity. What is required for being so is having similar domain of applicability and formalism.

4. Discuss the following statement about the equivalence principle: "According to the equivalence principle of general relativity, there is no experiment conducted in a small confined space that can distinguish between a uniform gravitational field and an equivalent uniform acceleration".
Answer: We note the following restrictions:
- Small confined space (what about large open space?).
- Uniform gravitational field (what about non-uniform gravitational field?).
- Uniform acceleration (what about non-uniform acceleration?).

Other extensions and generalizations of the equivalence principle can be similarly questioned.

3.2 Restrictions and Conditions on Lorentz Mechanics

1. What we mean by "speed" when we say: classical mechanics is a good approximation at low speeds? Also what we mean by "low"?
Answer: As explained before, the "speed" in this context means the relative speed between the frame of the observer and the frame of another observer or of the observed phenomenon, as symbolized typically by v in the expression of β and γ (i.e. $\beta = \frac{v}{c}$ and $\gamma = \frac{1}{\sqrt{1-\beta^2}}$),[30] while "low" means very small fraction of the characteristic speed

---

[30] In fact, v in the expression of β and γ is the main parameter, but not the only parameter, of speed in Lorentz mechanics as we will see in the discussion of velocity transformation and composition. In brief, for the validity of classical mechanics as a good approximation to Lorentz mechanics, the combined effect of any speed factor or factors that distinguish the formalism of Lorentz mechanics from the formalism of classical mechanics should become negligible compared to c and hence the Lorentz formulation will practically reduce to the classical formulation.

3.2 Restrictions and Conditions on Lorentz Mechanics

of light (i.e. $v \ll c$) that virtually makes $\beta \simeq 0$ and $\gamma \simeq 1$ for the intended purpose and this depends on the case and context. So, low speeds in one case or context could mean $v < 0.05c$ while in another case or context it could mean $v < 0.1c$.

2. Explain how the reduction of the Lorentzian formulation to the classical formulation at low speeds can be used as a validation test for the derived formulation of Lorentz mechanics.
   Answer: As discussed earlier, all the formulae of Lorentz mechanics should reduce to the formulae of classical mechanics when β tends to zero which is equivalent to very low speed v compared to c (i.e. $v \ll c$). Accordingly, this can be used as a test to validate the formalism of Lorentz mechanics where all the principal and subsidiary formulae of Lorentz mechanics should be validated by complying with this criterion, and hence any Lorentzian formula that does not pass this test should be rejected. This is based on the fact that classical mechanics is a trustworthy physical theory that has passed many tests in its domain of validity and hence if the postulated or derived Lorentz formulation is correct, it should converge to the classical mechanical limit at low speeds (i.e. $v \ll c$).

3. Analyze the expression that is used by some authors that classical mechanics is a good approximation to Lorentz mechanics (labeled as special relativity) when c tends to infinity.
   Answer: This expression should not be understood literally. In fact, $v \ll c$ and $c \to \infty$ are totally different conditions. What these authors mean is that if we consider c as infinite compared to our speed v (because the value of c is very large in comparison to the value of v) then classical mechanics is a good approximation to Lorentz mechanics. This is no more than another expression (which maybe distorted) for the condition that $v \ll c$ (or $\beta \simeq 0$ and $\gamma \simeq 1$) and hence it should not be understood literally. We should also express our reservation on the use of c in this context since c should be used as a symbol for the characteristic speed of light which is a constant and hence $c \to \infty$ seems meaningless.

4. Assess the "sacred" rule of modern physics that c is a speed restricted to light and it is the ultimate speed for any physical object.
   Answer: The proclaimed generality of this rule is based on the special relativity theory and its epistemological framework. This sacred rule is based on the sacred state of light in this theory. However, if the speed of light (although at a lower value than c) can be reached and exceeded in material media (which no one seems to dispute) then there is nothing in principle that prevents this from happening in free space. The current formulation of Lorentz mechanics that is restricted in its validity to this condition does not mean this is a universal fact but it means this is a condition that should be observed within the domain of validity of this mechanics and according to its current formulation.[31] So, outside the domain of validity of Lorentz mechanics or under certain extensions and modifications to this mechanics that could take place in

---

[31] Hence, we may say: this is a restriction on the current formalism rather than the speed, i.e. the current formalism is not valid beyond this restriction (due to a deficiency in this formalism) but the physical speeds beyond this restriction are not necessarily impossible. Accordingly, this restriction can be lifted by an extended or novel theory.

the future, it may be possible to reach and even exceed the characteristic speed of light c by some physical entities. In brief, having physical speeds $v \ge c$ should not be excluded automatically and dogmatically because of the requirement of a particular theory and its epistemological framework or because of the restriction on the Lorentz factor in the current formulation of Lorentz mechanics which can be lifted or modified in the future.

3.3 Space Coordination

1. Why space coordination is necessary in mechanics?
   Answer: Because mechanics is essentially the science of describing motion and following its development in space and time. So, a coordinate system is needed to identify the trajectory of motion and describe the development of physical phenomena in space and time.
2. List the main requirements of a coordinate system. What is the purpose of these requirements?
   Answer: The main requirements are:
   • The number of coordinates of the system should be equal to the number of dimensions of the space.
   • The coordinates should be mutually independent.
   The purpose of these requirements is to ensure that the employed space coordination is thorough and unique so that each point in the space is uniquely identified by a single set of coordinates, i.e. no point has no identification, no point has more than one identification, and each identification belongs to a single point and hence it is not shared with another point.
3. What will happen if the requirements of the previous question are violated?
   Answer: We investigate in the following points what will happen in the main cases of violation:
   • If the number of coordinates is less than the number of dimensions of the space then the coordination will not be thorough, i.e. the coordination will identify a subspace of the space rather than the whole space.
   • If the number of coordinates is equal to the number of dimensions of the space but these coordinates are not mutually independent then this will be equivalent to the previous case and hence the coordination will not be thorough.
   • If the number of coordinates is more than the number of dimensions of the space and these coordinates are mutually independent then we will have redundant coordinates that belong to a super-space and hence the coordination is not unique since each point in the space will have more than one set of coordinates due to the presence of extra coordinates which are free parameters. For example, if we employ three mutually independent coordinates in the coordination of the xy plane then the third coordinate is a free parameter and hence each point in the plane will be given the identification (x, y, z) with z being a free parameter that can take multiple values.
   • If the number of coordinates is more than the number of dimensions of the space and these coordinates are not mutually independent then this could result in having

non-unique identification, or redundancy in the set of coordinates, or even having non-thorough coordination.
4. What are the advantages of using rectangular Cartesian coordinate systems in space coordination?
   Answer: The main advantages of using rectangular Cartesian coordinate systems in space coordination are:
   • They are intuitive and hence they are easy to understand and manage.
   • They usually result in more simple mathematical formulation and manipulation.
5. What is the justification of the common use of rectangular Cartesian coordinate systems in the setting and formulation of Lorentz mechanics?
   Answer: The justification is that the space of Lorentz mechanics is flat and hence the rectangular Cartesian coordinate systems, which are intuitive and easy to use, are sufficient for achieving the required objectives.

3.4 Time Measurement and Synchronization

1. Try to make a deep analysis and criticism to the concept of synchronization.
   Answer: The definition of synchronization is, in fact, rather ambiguous and may even be circular. When we say "read the same time at any instant" what we mean by this particular instant as long as the observer in practice cannot be present in more than one location at any "given instant"? We believe this is deeply based on a hypothetical "global observer" who can observe all the times at all the points of space simultaneously. This may also necessitate not only the concept of global or universal time in the frame but even the concept of absolute time across all frames. No synchronization procedure can break the barrier of the confined presence of any real observer to make the time universal so that the synchronization concept becomes sensible and unambiguous. Anyway, we can regard these synchronization procedures as conventions for the definition of the concept of synchronization which accordingly is another convention that relies on a number of assumptions most of which are of philosophical rather than scientific nature. The reader can contemplate more about these issues which are fundamental to philosophy and science.
2. Discuss the validity of the first method of time synchronization in classical mechanics and in Lorentz mechanics.
   Answer: In classical mechanics, time flow is independent of the motion of the frame and hence all we need to ensure is that the time counting mechanism of the clock is not affected inadvertently by the motion (whether uniform or accelerated) due for example to mechanical vibration. We should also presume other assumptions such as the time counting mechanism of the clock is independent of the location in space (due for example to different ambient physical conditions) and the time counting mechanism of all clocks counts at the same rate. All these assumptions (whose essence is the uniformity of time counting) should be obvious and acceptable in classical mechanics and in general. The method should also need to be based on assuming that the synchronization procedure belongs to an inertial frame where the laws of physics (which are needed to justify and

3.4 Time Measurement and Synchronization   74

validate the synchronization procedure) are known to apply.[32]

Regarding Lorentz mechanics, the time in a given frame is potentially dependent on the movement of the frame regardless of the nature of this movement (e.g. relative to an absolute frame or relative to the frame of observation) and regardless of the nature of time dilation effect (e.g. real or apparent and in what sense). Accordingly, the first time synchronization method may seem difficult to justify. Yes, if we assume that the time counting mechanism is independent of the flow of time in the frame of mechanism and it can be fully controlled regardless of uniform and accelerated motion then this method can be justified. However, the common opinion is that all the physical processes (including all types of time counting mechanism) will follow the flow of time and hence they will slow down by the same rate of the time dilation effect. This means that it is not possible to find a physical process whose physical clock is independent of time flow where this process can be used in a time counting mechanism that universally applies to all clocks in the frame with the master clock at the origin of coordinates (which will stay at rest) providing a universal reference clock over the whole frame and for all clocks. But in our view it is still possible to regulate this physical clock (i.e. non-master) in a way to make it read time in a "wrong" way (e.g. by regulating its counting rate to be correlated with its speed) so that it always reads an "apparent" time that keeps in pace with the master clock, which is at rest, rather than with the time of the moving frame. In brief, as long as we can assume that we can find a clock that can keep in pace with the time of a frame other than its own frame, the first method of time synchronization should also be valid in Lorentz mechanics. We believe that even if regulating the time rate of basic physical processes (like atomic transitions in what we can call intrinsic time) to be out of pace with the time of the frame is impossible, it is possible to create and use synthetic (or composite) time counting mechanisms (like the mechanisms used in ordinary clocks) that can run out of pace with the time of the frame. This should be obvious when we see that it is easy for us to make a clock that runs twice the rate of ordinary clocks which "accurately" measure time flow. Such a synthetic mechanism can then be easily automated and regulated to have a count rate that potentially depends on the speed of the frame relative to the rest frame of the master clock so that both clocks will remain synchronized during the motion.[33] We should also note that not all fundamental physical processes should necessarily be affected by motion unless they depend on or correlated to the concept[34] that is used for defining, calibrating and measuring time, as indicated before and will be discussed further next.[35]

---

[32] To be more accurate and general, the condition should be: "the laws on which the synchronization procedure is based are known to apply in the synchronized frame" regardless of being inertial or not. However, in Lorentz mechanics inertial frames is the focus of interest and this should justify the restriction to inertial frames.

[33] In fact, the purpose of this potential dependency of the counting mechanism on speed is to correct for the dependency of the time flow in the frame of mechanism on speed so that the mechanism will ultimately become independent of speed.

[34] This concept (according to Lorentz mechanics) is the speed of light.

[35] We should remark that creating a faulty clock that follows the time flow of another frame is easy to imagine if the synchronized clock is adjusted continuously (manually or automatically) to be in pace

3.4 Time Measurement and Synchronization

3. Make a more fundamental distinction between the physical processes whose time rate should be intrinsically affected by the motion and those which should not.

   Answer: Logically, the time rate of any physical process that is based on or related to the physical concept that is used to define, calibrate and measure time should be intrinsically affected by the process of the defining concept, while the other physical processes should not. Now, since the definition, calibration and measurement of time (and space) in Lorentz mechanics are based on the speed of light, then the time rate of any physical process that is based on or related to the speed of light should be intrinsically affected by the motion (since motion affects time flow according to the defining concept of Lorentz mechanics), while the other physical processes should not. Accordingly, even the basic physical processes (let alone synthetic processes) that are not based on or related to the defining concept should not be affected by the factors that affect the process of the defining concept.

4. Apart from the universality of the speed of the employed signal within the given frame, there is another fundamental condition for the validity of the Poincare synchronization procedure. What is this?

   Answer: It is the inertiality of the synchronized frame (i.e. being inertial) since non-inertial frames are not guaranteed to follow the rules of physics which the synchronization procedure is based upon. In fact, this condition should apply to any synchronization procedure not only to Poincare procedure since any procedure is based on a number of laws and principles of physics that are needed to validate the procedure within the framework of the given mechanics.

5. Discuss if the invariance of the speed of light across all inertial frames (i.e. the second postulate of special relativity) is needed in the Poincare time synchronization procedure.

   Answer: The second postulate of special relativity is not required for the validity of the Poincare time synchronization procedure. What is needed is the universality[36] of the speed of light all over the given frame. Hence, even if the speed of light is frame dependent (i.e. not invariant across frames), the procedure is still valid as long as the speed of light is universal in each frame although it is not invariant across different frames.[37] Accordingly, light has no special status or particular significance in the Poincare procedure and hence any type of synchronization signal (e.g. sound) can be

---

with the clocks of another frame that is in relative motion with the frame of the synchronized clock (e.g. by continuously watching the clocks in the other frame and updating the synchronized clock accordingly). However, although this may be useful for demonstrating the possibility of creating such a faulty clock, it may not be useful for the purpose of synchronization since the clocks in the other frame should already be synchronized and hence we need a prime synchronization method that does not depend on a preceding synchronization procedure.

[36] Universality here means invariance (or constancy) of speed over a given frame (i.e. being independent of direction, location, time, etc.). We prefer to use universality instead of invariance to distinguish it from the invariance across different frames and hence avoid possible confusion.

[37] This may be envisaged, for example, in a classical projectile propagation model (not classical wave propagation model) with the validity of the Galilean transformations where the speed of light is universal in each frame (since it takes its characteristic value due to the relative rest between the clocks) although it varies across different frames since it transforms by the Galilean transformations.

3.4 Time Measurement and Synchronization

used in this procedure as long as the condition of the universality of speed over the given frame and the condition of being in an inertial frame are satisfied. In fact, even the condition of universality of speed over the given frame is not needed if we assume that a correction mechanism, for the dependency on direction for example, can be found to account for the dependency of speed in the given frame.

6. Assess a Poincare-like time synchronization procedure that is based on the use of sound, instead of light, where the source of sound is in the rest frame of the medium of propagation, e.g. air or water.
   Answer: Since the source of sound (i.e. the master clock at the origin of coordinates) is at rest with respect to the medium of propagation (which should also be at rest with respect to this coordinate system), then there should be no direction dependency (at least due to the motion of medium) of the speed of synchronization signal and hence the procedure should be as valid as the procedure that is based on using light signal. Obviously, the other assumptions about the constancy of the speed of the synchronization signal over the whole frame regardless of location, time, etc.[38] similarly apply for the same reasons. The condition of being in an inertial frame should also be assumed in this synchronization procedure to ensure that the laws of physics that underlie the synchronization procedure hold true in the synchronized frame.[39]

7. Is the Poincare method of synchronization proprietary to Lorentz mechanics or it can also be used in classical mechanics?
   Answer: As long as the above conditions of universality of speed and inertiality of frame are met, the method is valid in classical mechanics as in Lorentz mechanics and hence it is not proprietary to Lorentz mechanics.

8. Discuss the validity of the Poincare time synchronization procedure according to classical mechanics.
   Answer: If we adopt the view of classical mechanics that the characteristic speed of light c belongs to a privileged frame (whether absolute space or ether) then in any frame that is not at rest with respect to the absolute frame the speed of light will be dependent on the direction of propagation and hence the condition of universality of speed will be violated.[40] Yes, a modified version of this procedure in which the dependence on direction is taken in the calculation of the offset time can still be used although it will be more complicated and may be practically prohibitive even if it is fundamentally viable.
   We should remark that the proposition of this time synchronization method by Poincare

---

[38] This should also include potential direction dependency due to factors other than the motion of medium such as the presence of external fields or inhomogeneity or anisotropy of medium.

[39] As indicated before, the important thing is that the laws which the synchronization procedure depends upon are guaranteed to hold in the synchronized frame. Being inertial is simply inline with the fact that we are dealing with Lorentz mechanics where the frames should be inertial and hence the laws are required to hold in inertial frames.

[40] As indicated earlier, this applies for a classical wave model for the propagation of light (and this should be inferred from "privileged frame"). Accordingly the Poincare time synchronization procedure should be classically valid if we adopt a projectile propagation model because the source and receiver are in relative rest.

should suggest his rejection to the classical Galilean models for the propagation of light[41] and this (among other indications that we suggested earlier) could be the origin and basis for the second postulate of special relativity (although the procedure in itself does not require this postulate according to our analysis and consideration of various propagation models).

9. A clock that is at $\mathbf{r}_c = (0.1a, -1.3a, 8.4a)$ where $a = 3 \times 10^8$ is to be synchronized by the Poincare procedure. If the master clock is at $\mathbf{r}_m = (-0.35a, 3.4a, 0.4a)$ what is the offset time of the clock assuming that the speed of light in that frame is c.
   Answer: The distance d between the two clocks is given by:
   $$\begin{aligned} d &= \sqrt{(\Delta x)^2 + (\Delta y)^2 + (\Delta z)^2} \\ &= a\sqrt{(0.1+0.35)^2 + (-1.3-3.4)^2 + (8.4-0.4)^2} \\ &\simeq 9.2894a \end{aligned}$$
   Hence, the offset time should be:
   $$\Delta t = \frac{d}{c} \simeq 9.2894\,\mathrm{s}$$

10. What sort of symmetry the offset time of clocks in the Poincare procedure should have? Also, where is the location of the center of symmetry?
    Answer: The offset time of the clocks should have spherical symmetry centered at the master clock because it is assumed that the light speed is the same in all directions and hence the light signal, which is emitted by the master clock, will reach at the same time to all the points on the surface of any sphere of a given radius with that center. This situation of spherical symmetry will be violated if the speed is not the same in all directions and hence a modified version of the Poincare procedure, that is based on using a method to correct for direction-dependency, is employed.

3.5 Calibration of Space and Time Measurement

3.6 Reference Frame

3.6.1 Construction of Reference Frame

1. Describe in detail how to construct a simple and tidy frame of reference using the Poincare synchronization procedure.
   Answer: To construct a frame of reference we follow the following procedure assuming that we have access to length- and time-measuring equipment (e.g. measuring sticks and clocks):[42]

---

[41] At least the wave propagation model; the projectile propagation model can then be excluded because it was not considered as a possibility by the overwhelming majority of the scientific community at that time due to the overwhelming acceptance of the wave propagation model.
[42] We should also need equipment to identify directions (or angles) in space.

3.6.1 Construction of Reference Frame 78

- We start by introducing a 3D rectangular Cartesian coordinate system that envelops the whole space.
- We then introduce to the space a 3D array of clocks positioned equidistantly in each one of the three spatial directions. These clocks are normally assumed to be evenly-spaced in each direction although the spacing does not need to be the same in all directions (e.g. 1 meter spacing in the x-direction, 1.5 meter in the y-direction and 0.75 meter in the z-direction) or even in any direction. Although being evenly spaced is not a necessary condition, it is needed if we want to have a tidy reference frame. These clocks, which are initially in a state of stand still, have an automatic initialization mechanism that makes them start running as soon as they receive a light signal from the origin of coordinates.[43]
- The master clock at the origin of coordinates is set to read zero time (00:00) while each one of the other clocks is set to read the time required for a light signal sent from the origin of coordinates to reach that clock.[44]
- A light signal is then sent from the origin of coordinates in all directions with the start of the master clock at the origin. As soon as any one of the other clocks receives the light signal, it starts running and because of the time offset of that clock, its reading will be identical to the reading of the master clock at the origin. Hence, eventually all these clocks in this 3D array will read the same time at any instant.

By following this procedure, we create a reference frame that can be used to assign unique and well defined spatial and temporal coordinates to any event that occurs in the event space. We can also identify the world line of any object or correlated series of events.

2. Discuss and assess the frame construction procedure that you described in the previous exercise.

   Answer: Considering the answer that we provided in the previous exercise, we note the following points:
   - The spatial coordinate system of the frame can be non-Cartesian, as well as Cartesian, which is a simple generalization to the above-described procedure. We chose in the answer a Cartesian system to have a simple frame.
   - The described construction procedure is based (following the instructions of the previous exercise) on using Poincare synchronization method, as discussed in the text. This method is the one that is commonly used in the literature of Lorentz mechanics. As discussed earlier, the validity of this synchronization method is based on the assumption of the universality of the speed of light within that frame (or on the assumption of viability of employing direction-dependency corrections)[45] and on the assumption of

---

[43] The purpose of this elaborate description (as if the construction is a real process) is to have precise and tidy concepts and procedures. This should also help in training on organized thinking and developing methodical ideas. This may also help in demonstrating the practicality of constructing a reference frame and ruling out any doubt or skepticism about its physical sensibility and viability.

[44] In fact, what is really needed is that the clocks are set with the above offset time difference regardless of the initial time of the master clock as being zero or not. We also note that the master clock is not required to be at the origin of coordinates although this will make the procedure more tidy and simple.

[45] In fact, if we follow the above statement (i.e. "each one of the other clocks is set to read the time

inertiality of the frame.
- Other synchronization methods can also be used in the frame construction procedure if their working principle is accepted and validated. The construction procedure should then be modified to accommodate such changes in the synchronization method.

3. Explain how an observer can assign spacetime coordinates to events that occur in a given frame.
  Answer: With the above-described reference frame (refer to the answers of the previous exercises), an observer (whether in that frame or in another frame) can assign spacetime coordinates to each event in the given frame by simply recording the spatial coordinates of the array point nearest to the event and reading the clock placed at that point. Any degree of accuracy can be achieved by employing a sufficiently resolved densely-packed array to fulfill the required accuracy. We note that although the observer can be in that frame or in another frame, the assigned spacetime coordinates belong to the given frame, i.e. not to the frame of the observer unless the observer is in the given frame.

3.6.2 Inertial and non-Inertial Frames

1. Define inertial frame in a memorable way.
  Answer: *Inertial* frame is a frame of reference in which the law of *inertia* holds true. This definition sheds light on the origin of this label.
2. Compare between the procedural definition (i.e. the one based on Newton's laws) and the fundamental definition (i.e. the one based on the state of motion relative to absolute frame) of inertial frame.
  Answer: Some valid comparison points are:
  - The procedural is purely physical (or scientific) while the fundamental is not so because it has epistemological content by referring to the absolute frame.[46]
  - The procedural can be circular or redundant if the objective is to identify the frames in which the laws of classical mechanics hold true while the fundamental cannot.
  - The procedural is based on the symptoms while the fundamental is based on the cause.

  Accordingly, the procedural may be seen as superior from the perspective of the first point while the fundamental may be seen as superior from the perspective of the second and third points.
3. Characterize inertial and non-inertial frames in terms of their velocity and acceleration in space.
  Answer: Inertial frames are characterized by having constant translational velocity (i.e. constant in magnitude and direction) in space. Accordingly, non-inertial frames

---

required for a light signal sent from the origin of coordinates to reach that clock") then this condition is redundant. However, we stated it in this way assuming a simple formula like $t_o = d/s$ is used to calculate the offset time. This may also apply (partially at least) to the condition of inertiality.

[46] In fact, absolute frame may be defined physically (rather than epistemologically) as the frame that is based on the actual distribution of matter and energy in the Universe on astronomical and cosmological scales. However, this should be based on actual observational and experimental evidence which may be difficult to establish although it should be valid as an approximation at least.

are characterized by having variable velocity. This may arise because of variation in magnitude or in direction or in both. In terms of acceleration, the acceleration of inertial frames is zero while the acceleration of non-inertial frames is non-zero. The acceleration can be translational or/and rotational where each can be constant or variable. We therefore have translational acceleration where the accelerated motion is taking place in a fixed direction and rotational acceleration where the accelerated motion is taking place with a constant translational speed.[47] We may also have a more general type of acceleration where both the magnitude and direction of velocity vary. We note that all forms of rotational motion relative to an inertial frame represent a state of acceleration, while only non-uniform translational motions represent a state of acceleration.

4. Assess and criticize the use of concepts like velocity and acceleration in the definition of inertial and non-inertial frames where these concepts may require a "master reference frame" to which these concepts are referred.

   Answer: From the answer to the previous question, we see that we need an underlying spacetime if the definition and characterization of inertial and non-inertial frames should have any meaning because having constant or variable velocity or being accelerated or non-accelerated requires a space with well defined length scale, time scale and directions so that having constant or variable velocity or zero or non-zero acceleration have sensible meaning. This leads us to the issue of absolute space and absolute time where the physical difference (which is real and observable) between inertial frames that hold Newton's laws and non-inertial frames that do not hold Newton's laws is difficult to explain and justify in the absence of absolute space and absolute time (or absolute frame) because real physical effects cannot originate from differences that are based on our definitions and conventions where we arbitrarily consider one frame to be inertial because it is in uniform motion with respect to an arbitrarily selected frame while we arbitrarily consider another frame as non-inertial because it is accelerating with respect to an arbitrarily selected frame. In fact, basing our definitions on such arbitrary choices should lead to contradictions because the same frame can be inertial and non-inertial when it is referred to different frames, and this obviously violates the rules of reality and truth.

5. Show that a frame which is in a state of acceleration relative to a non-inertial frame can be inertial or non-inertial and hence its inertiality cannot be easily identified from simple observations. Demonstrate this graphically.

   Answer: Let demonstrate this by an example. If one frame O_1 is moving (say relative to absolute space in a 1D motion along the x direction) with an equation of motion $x_1 = 10t$ and another frame O_2 is moving with an equation of motion $x_2 = 5t^2$ then they are in a state of relative acceleration where the first frame is inertial while the second frame is not. But if we have a third frame O_3 that is moving with an equation of motion $x_3 = t^3$ then it is in a state of relative acceleration with respect to both O_1 and O_2 (since the second time derivative of its velocity is not vanishing while the second

---

[47] This type of accelerated motion by having constant speed with changing direction is more general than rotational motion around a circle or around a regular helix and hence it should be understood in this sense.

3.6.2 Inertial and non-Inertial Frames

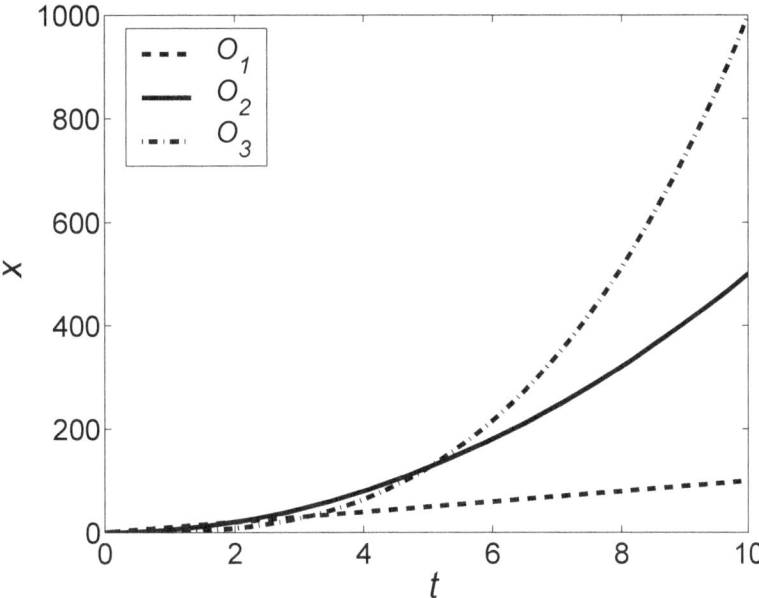

Figure 3: The plot of x versus t for the three frames (O_1, O_2 and O_3) of exercise 5 where it is seen that the slopes of O_1 and O_3 relative to the slope of O_2 are varying (i.e. time dependent) and hence O_1 and O_3 are seen by O_2 to be in a state of acceleration although O_1 is inertial while O_3 is non-inertial.

> time derivative of the velocity of O_1 and O_2 are both vanishing) and it is obviously accelerating or non-inertial. Hence, relative to O_2 both O_1 and O_3 are accelerating but while O_1 is inertial O_3 is non-inertial. Accordingly, the frame O_2 cannot determine by simple observations of the relative motion of O_1 and O_3 the state of these frames and if they are inertial or not. This example is graphically demonstrated in Figure 3 where we see the slopes (which represent the 1D velocity) of the curves of O_1 and O_3 varying in time relative to the slope of O_2 and hence both O_1 and O_3 are seen by O_2 as accelerating frames although O_1 is inertial while O_3 is non-inertial.

6. Use a graphical argument, similar to the one in the previous exercise, to show that the state of inertiality of any frame can be easily identified from simple observations from any given inertial frame.
   Answer: In Figure 4, we re-plot Figure 3 where we replace O_2 with O_4 which is an inertial frame whose equation of motion is given by $x_4 = 50t$. As we see, the slope of O_1 relative to the slope of O_4 is constant (since the angle between the tangent vectors of two straight lines is constant and hence the tangent of this angle which represents the relative velocity is also constant) while the slope of O_3 relative to the slope of O_4 is time-dependent variable. Accordingly, from frame O_4 (whose observer knows that his frame is inertial) the inertiality status of O_1 and O_3 can be easily determined by simple observations, i.e. by measuring the relative velocity to see if it is constant or variable.

7. Show, using a simple non-rigorous argument, that in the definition of inertial frames the condition of holding Newton's first law and the condition of holding all three Newton's

3.6.2 Inertial and non-Inertial Frames

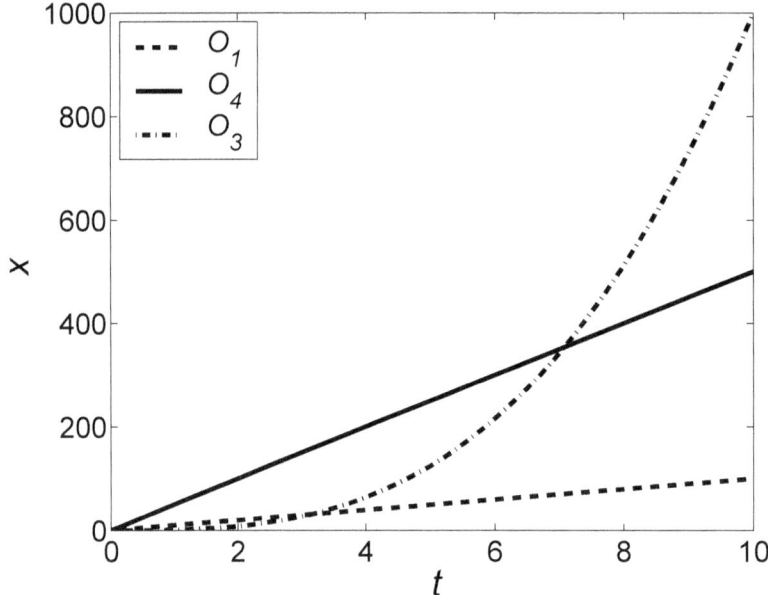

Figure 4: The plot of x versus t for the three frames (O_1, O_4 and O_3) of exercise 6 where it is seen that the slope of O_1 relative to the slope of O_4 is constant in time and hence O_4 can conclude from simple observations that O_1 is inertial, while the slope of O_3 relative to the slope of O_4 is varying in time and hence O_4 can conclude from simple observations that O_3 is non-inertial.

laws are equivalent.

Answer: It is obvious that if the three Newton's laws of motion hold true then Newton's first law should also hold true because it is one of these laws. We need then to show that if Newton's first law holds true then the second and third laws also hold true. So, let assume that Newton's first law holds true in an inertial frame but Newton's second law does not hold true. Now, if we interpret Newton's second law as a statement of the fact that force is a necessary and sufficient condition for accelerated motion then these two premises (i.e. "Newton's first law holds true" and "Newton's second law does not hold true") are obviously inconsistent because Newton's first law means that force is a necessary and sufficient condition for acceleration and hence in the absence of force there is no acceleration (and vice versa) as stated by this law.[48] Regarding Newton's third law, it is less obvious that if Newton's first law holds true then the third law should also hold true. However, we can argue that if we interpret Newton's second law this time as a statement about characterizing the nature of force, then if Newton's third law does not hold true (and hence the symmetry of forces between the two interacting

---

[48] We note that the premise "force is a necessary and sufficient condition for accelerated motion" which is attributed to both Newton's first and second laws does not mean that the two laws are the same since the second law is distinguished by the quantification of force and its relation to momentum where this quantification is not involved in this comparison between the two laws. However, this could be a valid challenge to this argument.

objects is broken) then the force as characterized by Newton's second law is not well defined and this means a violation to Newton's second law as a law.[49] Now, since Newton's first law is a special case of Newton's second law then this also implies a violation to Newton's first law.[50] Hence we cannot have a situation in which Newton's first law holds true but Newton's second law or third law does not hold true. We should emphasize that this argument is provided for pedagogical purposes and hence it is not rigorous or conclusive and can be easily challenged. Anyway, the implication of Newton's first law to Newton's second and third laws can be established as an empirical fact based on experiment and observation, i.e. any frame in which Newton's first law is observed to hold true Newton's second and third laws are also observed to hold true.

8. A camera is installed on a rotating platform, and another camera is installed on the ground. Assess how the two cameras will record the events that occur in their surroundings.
Answer: The camera on the ground represents the eye of an inertial observer (where the Earth approximates an inertial frame) while the camera on the rotating platform represents the eye of a non-inertial observer. For example, a ball moving freely on a flat ground will be seen by the platform camera in a state of accelerated motion (since it follows a curved path) but with no external force, while it will be seen by the ground camera in a state of uniform motion. Accordingly, Newton's laws are seen invalid by the platform camera and valid by the ground camera.

3.6.3 Difference between Inertial and non-Inertial Frames

1. Discuss the necessity for the existence of an absolute frame for the reality of the physical difference between uniformly moving frames and accelerating frames as being inertial and non-inertial.
Answer: From observing the physical world, we see that we have two categories of reference frames: a category in which Newton's laws hold and another category in which these laws do not hold. We also see that all the frames in the first category are characterized by being in a state of uniform motion relative to each other while all the frames in the second category are characterized by being in a state of acceleration with respect to the frames in the first category. As the frames in each category vary in all aspects, and hence from conventional viewpoint they are not distinguished from the frames in the other category more than they are distinguished from each other, then the only logical explanation is that the frames in the first category are in a state of uniform motion with respect to an absolute frame (and hence they are in uniform motion with respect to each other) while the frames in the second category are accelerating with respect to this absolute frame (and hence they are accelerating with respect to the frames in the first category) and that is why the two categories differ and the difference

---
[49] This part of the argument may also be established tentatively by the link between the second and third laws and the conservation of momentum as an established fact although this logic may also be challenged.
[50] This can also be challenged.

3.7 Causal Relations 84

in their physical characteristics is real.
2. Discuss the suggestion that what discriminates between inertial and non-inertial frames is not the existence of absolute space (or rather absolute frame) but the actual large scale distribution of matter and energy in our space that produces these different effects and makes some frames behave as inertial while others as non-inertial.
Answer: In our view, this does not make any real difference. The suggestion should be regarded as a diversion from the important physical question to a distracting philosophical contemplation. What we mean by absolute space (or absolute frame) is the space that we live in, i.e. the space of the physical world whose laws and rules are the primary interest for science. Whether the properties of this space arise because of the intrinsic nature of this space or because of the large scale distribution of matter and energy inside this space is irrelevant.[51] What is important to physics is the actual space (not an abstract hypothetical space) and its ability to provide an absolute frame of reference to justify the physical difference between inertial and non-inertial frames which is real and observable.

3.7 Causal Relations

1. Assess Newton's law of gravitation and Coulomb's law of electrostatic force in the light of the concept of action at a distance.
Answer: Despite the criticism to these laws that they presume or imply action at a distance, since the speed of communication between masses or charges should be infinite in the absence of any temporal parameter in the equations of these laws, they are not so. In fact, they principally can represent steady state situations following a communication during transient states. They can also represent interaction between objects that are separated by rather short distances so practically all communications appear as if the speed of signals is infinite. It is obvious that in our answer we are assuming that such communications by signals of finite speed are required to establish causal relations to avoid the allegedly impossible action at a distance. We note that this issue should also be linked to other issues, like the issue of quantum non-locality which is a controversial issue that is thoroughly investigated in quantum mechanics, and hence the alleged impossibility of the presumed "action at a distance" should not be considered an undisputed fact.
2. Based on the nature of causal relations, what is the relation between two simultaneous events?
Answer: If we rule out the possibility of action at a distance, then two simultaneous events (refer to § 1.3 for definition) cannot be linked by a causal relation unless they are at the same location in space (i.e. they are co-positional as well and hence they are identical) because no signal can have an infinite speed so that it communicates between the two events in a zero time duration.[52]

---

[51] This should be true at least from a practical viewpoint because our space cannot be stripped of matter and energy.
[52] As indicated earlier, we need to sensibly envisage the type of communication signal (if it is needed at

3.8 Speed of Light

1. Discuss and assess the main possibilities about the speed of light in inertial frames of reference.

   Answer: There are several possibilities for the speed of light in inertial frames which are the domain of Lorentz mechanics:

   • The characteristic speed of light c belongs to a privileged absolute frame (say the absolute space or ether frame) and hence the observed speed of light for any inertial observer will be $c \pm v$ where v is the speed of the observer relative to the absolute frame and where $\pm$ signs are chosen according to whether c and v are in the same direction or in opposite directions, i.e. $c + v$ if the observer is moving towards the source of light and $c - v$ if the observer is moving away from the source of light. This possibility, which follows the model of material waves in classical mechanics, is generally (but not necessarily) based on the ether theory which assumes the existence of ether as a propagation medium for light whose frame may be identified with the frame of absolute space.

   • The characteristic speed of light c is relative to its source and hence the observed speed of light for any inertial observer depends on the relative motion between the source and observer, i.e. the speed of light as measured by an observer moving with a speed v relative to the source is $c \pm v$ where $+/-$ is used when the observer is moving towards/away from the source. Although this possibility, which treats light like material projectiles in classical mechanics (refer to § 1.13.1), does not need to assume an absolute frame, it is not based on denying the existence of absolute frame and hence it is neutral towards the possibility of the existence of an absolute frame.

   • The speed of light is frame independent and hence the observed speed of light takes the characteristic value c in all inertial frames, i.e. it is absolute and equally applies to all inertial observers. Now, although this possibility is normally associated with the theory of special relativity that denies the existence of absolute frame, the denial of the existence of absolute frame is not a logical requirement for this possibility. In fact, this potentially depends on the interpretation of being "universal" and "the same in all inertial frames" and whether this is real or apparent for example. So, if we interpret this universality in a certain sense then this can be consistent with the existence of absolute frame while it may not be so according to another interpretation (refer for example to § 1.6 and § 11).

   We note that while the first two possibilities are consistent with the Galilean rule of velocity composition (although the interpretation of v differs in these possibilities), the third possibility is not. Accordingly, the first two possibilities should apply only within the framework of classical mechanics while the third possibility applies within the framework of Lorentz mechanics. We would like to emphasize on the fact that while the first possibility logically implies the existence of absolute frame, since it is based on the existence of such a frame, the second and third possibilities are logically consistent

---

all) when the events are co-positional or identical. This may cast a shadow at least on the generality of the requirement of communication signal for establishing causal relations.

3.8.1 Speed of Light as Restricted and Ultimate Speed

with both choices, i.e. existence and non-existence of absolute frame.[53]

3.8.1 Speed of Light as Restricted and Ultimate Speed

1. Summarize the stand of special relativity about the limits and restrictions on the speed of physical objects.
   Answer: According to special relativity, no massive object can reach or exceed the characteristic speed of light c. Moreover, all massless objects, such as light and other forms of electromagnetic radiation, must travel at this characteristic speed c.
2. Why we insist on the claim that reaching and exceeding the characteristic speed of light c is possible in principle if it is denied by the formalism of Lorentz mechanics?
   Answer: Our view about allowing this possibility despite the apparent clash with the current formalism of Lorentz mechanics is justified in part by the many claims of observing superluminal speeds in experimental and astronomical studies and these claims should be given the opportunity for in-depth examination. Our view is more justified with regard to non-inertial frames where there is no restriction from the formalism of Lorentz mechanics due to the invalidity of this mechanics in these frames and hence its restrictions do not apply.
3. What are the main restrictions that should be imposed on the claim that no massive object can reach or exceed the speed of light?
   Answer: The following restrictions should be imposed on this claim:
   • This claim, if established, applies only to free space. Accordingly, massive particles can travel faster than light in material media such as water, and this is the basis for some interesting physical phenomena such as Cherenkov radiation.
   • This claim, if established, is within the domain of Lorentz mechanics, and hence it applies only to inertial frames. Any extension to non-inertial frames requires further evidence.
   • Even within the domain of Lorentz mechanics, this claim, if established, should be restricted to the current formalism of Lorentz mechanics. This is because the establishment of this claim within the domain of Lorentz mechanics by the above arguments (e.g. singularity of Lorentz factor when $v = c$ or becoming imaginary when $v > c$) or any similar arguments does not necessarily mean the establishment of this claim as an absolute fact because Lorentz mechanics can be a special theory that can represent a limiting or special case for a more general theory that can be developed in the future. In brief, we should allow for the possibility that Lorenz mechanics in its current formalism can be a restricted theory or an approximation to a more general theory as we found in the past that classical mechanics (which was once believed to be a general and final theory before the collapse of this belief with the emergence of modern physics) is not a general theory.
4. Explain how the lateral speed of light can exceed the constant c.
   Answer: Let have a rotating light beam where the light source is at the center of a circular screen that surrounds the light source (refer to Figure 5). The lateral speed of

---

[53] However, the interpretation of the third possibility may casually introduce a logical inconsistency.

3.8.1 Speed of Light as Restricted and Ultimate Speed

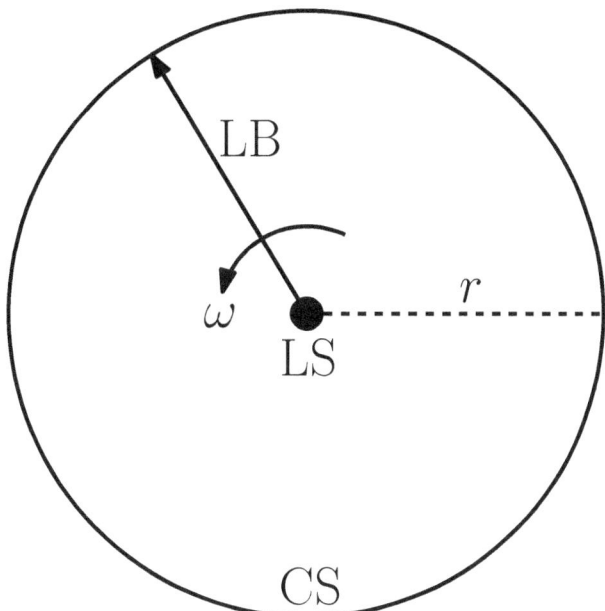

Figure 5: A light source LS at the center of a circular screen CS with radius r where the source is rotating at angular speed ω and casting a light beam LB on the screen (in fact the beam LB should not be straight but it is drawn straight for simplicity).

the projection of the light beam on the screen is given by:

$$v = \omega r$$

where ω is the angular speed of the light source and r is the radius of the screen. Now, if we want v to be superluminal (i.e. $v > c$) then the following condition should apply:

$$\omega r > c \quad \Rightarrow \quad \omega > \frac{c}{r}$$

For example, if the radius r is numerically equal to $0.01c$ then the angular speed should be $\omega > 100$ (i.e. about 16 revolutions per second) which is a mechanically viable speed (also see § 10.4).

5. It has been shown in the literature of special relativity that the speed restrictions are required for preserving causal relations between events. Can this be regarded as a proof for these restrictions?[54]

Answer: No. Assuming that all the technicalities of this "proof" are correct (some of which at least are certainly arguable especially those based on graphic demonstrations like light cone), it is based in the first place on the theoretical framework of Lorentz mechanics and hence within another theoretical framework (as a result of amendment or replacement of the current formalism of Lorentz mechanics) and outside its validity

---

[54] It is claimed in the literature of special relativity that the desire to preserve causal relations is the basis for c being regarded as an ultimate speed.

domain it is quite possible for these restrictions to be lifted. We also note that it may be claimed that the formalism of Lorentz mechanics is based on the speed of light and hence even if these restrictions are established (by this argument or another argument) they should be related to this speed and not to any other speed of physical objects especially massless objects. More fundamentally, the premise that causal relations must be established by communicating physical signals that are subject to the speed restrictions of special relativity can be challenged (see § 3.7). The reader is referred, for example, to the literature of quantum mechanics about non-local reality of some quantum phenomena.

3.8.2 Measuring the Speed of Light

1. Assess the methods of one-way and two-way measurement of the speed of light considering the issue of possible dependency of this speed on direction.
   Answer: For the one-way methods (if assumed possible), the measurement can be conducted in one way then the other way and hence a conclusion about the dependency of the speed of light on direction, which is a direct test to the frame dependency and the Galilean composition rule, can be reached. This may seem impossible in the two-way methods where only the average speed in the two directions can be measured. However, the potential dependency on direction can still be examined by rotating the device similar to the procedure of the Michelson-Morley experiment (refer to § 2.6 and § 12.2) although we do not need to have interference between two beams. In fact, even the nature of the orientation dependency, the orientation of the relative speed and the value of the relative speed (assuming it is constant) can be concluded from examining the dependency of speed in different orientations if we follow the method of Michelson-Morley experiment and accept the working principles of its analysis (refer to the next exercise) which include adopting a wave propagation model (and hence a different approach is needed if a projectile propagation model, for example, is assumed).
2. A scientist is trying to measure the "ether wind" that is caused by the Earth movement in space. He uses a device that measures the time of flight of light in different orientations by rotating the device and recording the time as a function of orientation. The device simply measures the time of the two-way trip of light over a length $L = 15$ meter in its forward-backward journey. The scientist observed a maximum two-way trip time of 1.0001×10^{-7} second in the east-west orientation and a minimum two-way trip time of 1.00005×10^{-7} second in the north-south orientation. Assume the validity of the Galilean velocity composition rule and use the Michelson-Morley analysis method (refer to § 2.6 and § 12.2) which is based on classical mechanics to infer the speed of the ether wind and its orientation according to this method.
   Answer: According to the Michelson-Morley analysis the time of the two-way light trip in the parallel orientation to the ether wind is given by:
   $$t_1 = \frac{L}{c-v} + \frac{L}{c+v} = \frac{2cL}{c^2 - v^2} = \frac{2L\gamma^2}{c}$$
   while the time of the two-way light trip in the perpendicular orientation to the ether

wind is given by:
$$t_2 = \frac{2L}{\sqrt{c^2 - v^2}} = \frac{2L\gamma}{c}$$
Because $\gamma > 1$, it is obvious that $t_1 > t_2$, and hence the observed maximum time corresponds to the parallel orientation and the observed minimum time corresponds to the perpendicular orientation. On taking the ratio of t_1 to t_2 we obtain:
$$\frac{t_1}{t_2} = \frac{1.0001 \times 10^{-7}}{1.00005 \times 10^{-7}} \simeq 1.00005 = \gamma$$
and hence the speed of the ether wind is $v \simeq 0.01c$. From the above analysis, the maximum time of flight corresponds to the orientation that is parallel to the ether wind and hence the ether wind is in the east-west orientation.

We should remark that this method is based on a classical wave propagation model because its objective is to measure the "ether wind". Hence, if we want to test potential direction dependency of the speed of light in general, then we need to have a different approach that also considers propagation models other than the classical wave model (e.g. projectile model).

3.9 Spacetime in Lorentz Mechanics

1. Why we are forced to travel in time and only in forward direction but not in space? Does this break the symmetry (or equivalence) of the spacetime coordinates and hence space and time are still not completely merged in a spacetime manifold (which should usually have equivalent dimensions) even in Lorentz mechanics?
   Answer: Yes, even in Lorentz mechanics space and time are rather still distinct and the merge of these into spacetime is not complete. For example, we can travel freely in space but not in time. Also, we can travel forward and backward in space but we can travel only forward in time. This distinction is seen even in the mathematical formulation of the spacetime interval and the metric of Minkowski space where the spatial and temporal coordinates have opposite signs. We should also note that the merge of space and time (although in a lesser extent) can also be seen in classical mechanics where the spatial coordinate in the orientation of movement also contains a temporal factor according to the Galilean transformations (i.e. $x' = x - vt$). So, Lorentz mechanics is not entirely novel or complete in this regard.
2. Discuss the issue of invariance of ds, dt and $d\sigma$ in classical mechanics and in Lorentz mechanics and the implication of this on the underlying spaces of these mechanics.
   Answer: As indicated already in the text and will be discussed further later on, while ds and dt are separately invariant in classical mechanics but not their mix as represented by $d\sigma$, in Lorentz mechanics the opposite is true where $d\sigma$ is invariant but not ds or dt. This indicates that what underlies classical mechanics is two separate manifolds: a 1D temporal manifold and a 3D spatial manifold. In contrast, what underlies Lorentz mechanics is a single manifold which is the 4D spacetime manifold. This highlights the issue of the merge of space and time of classical mechanics into spacetime in Lorentz mechanics.

3.9.1 Spacetime Diagram and World Line

1. Define the concept of "event" in spacetime.
   Answer: An event in spacetime is an occurrence that is fully identified by three spatial coordinates and one temporal coordinate and hence it is represented by a single point in the 4D spacetime manifold (or Minkowski space).
2. Determine if the coordinates of spacetime, as represented by its "coordinate system", have the same physical dimension or not.
   Answer: Because the convention is to represent the temporal dimension of the spacetime by ct rather than t, all the coordinates of the spacetime "coordinate system" have the same physical dimension which is length.
3. Plot a simple 2D spacetime diagram on which you draw the world lines of the following objects: (a) An object that moves in time but not in space. (b) An object that moves in space but not in time. (c) An object that moves in space and in time. (d) An object that moves neither in space nor in time. Comment on these objects.
   Answer: A plot may look like Figure 6 where:
   (a) This is represented by the vertical line labeled a. This is physically possible. However, it should be impossible (according to the current state of physics) if the arrow is reversed and hence the object moves backward in time.
   (b) This is represented by the horizontal line labeled b. This is physically impossible (at least within the framework of Lorentz mechanics and the contemporary physics) because it requires infinite speed.
   (c) This is represented by the curve labeled c. This is physically possible in some parts and impossible in other parts. The parts that do not reach or exceed the speed of light or move backward in time should be unanimously physically possible while the other parts should be either out of reach of the formalism of Lorentz mechanics or physically impossible according to the contemporary theories of physics. This world line will be completely (or almost completely) impossible if we reverse the direction of traversing this path.
   (d) This is represented by the dot labeled d. This is physically impossible if it represents a continuously existing object since it does not move in time although it may represent an event that happened at an instant of time.
   Comment: from the results of this exercise, we can observe a fact about spacetime that is although Lorentz mechanics succeeded in mingling the space and time into a single spacetime manifold, they are still not totally equivalent because of this asymmetric situation where we see that an object can "stop" in space (since it has the freedom to choose its location) but it cannot "stop" in time since it has no freedom to choose its position in time because the flow of time is out of our control. So, even in Lorentz mechanics this mingling is not perfect or complete (future and even some current physical theories may improve and enhance this mingling). Another demonstration of asymmetry between space and time is that we can go in both directions (forward and backward) in space but we can only go forward in time; this may be based on the previous asymmetry (or it may be equivalent to it). In this regard, we may also note that while contraction

3.9.1 Spacetime Diagram and World Line 91

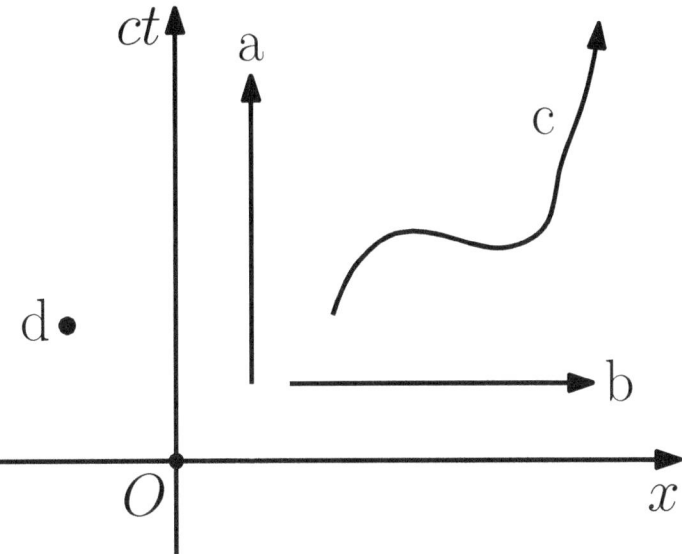

Figure 6: A 2D spacetime diagram on which we represent (a) the world line of an object that moves in time but not in space (at least in the x direction), (b) the world line of an object that moves in space but not in time, (c) the world line of an object that moves in space and in time, and (d) the "world line" of an object that moves neither in space nor in time.

>of spatial coordinates occurs only in the direction of motion, as we will discuss later, contraction of time coordinate occurs regardless of any direction since time is not supposed to be directional although it can be claimed that any motion should be in the direction of the temporal coordinate since any physical occurrence should take place in time.

4. Based on your answer to the previous question, determine if moving at the speed of light means stopping in time as claimed by some special relativists.
   Answer: It should not be so (although it may look like this based on the current formulation of Lorentz mechanics). The reason should be obvious because the world line of a light signal is represented by the straight line $x = ct$ (or $x^1 = x^0$) which is not a horizontal line and hence as long as there is a change in the space coordinate (which is obvious since light is moving in space) there should be a movement in time and hence a light signal will be detected at one location at a given time and at a different location in a later time. So, an object will stop in time only if it follows a horizontal line (or stay on a single point in spacetime which can be considered as a special case of horizontal line). However, stopping in time according to the above claim is based on the time dilation effect where time seems to stop as the moving object approaches the speed of light c. In brief, we should differentiate between the frames (i.e. frame of observer and frame of observed) in considering this claimed stopping in time.

5. Briefly define world line.
   Answer: It is a path or trajectory in the spacetime or it is the representation (by a

line or curve) of this path on the spacetime diagram. Accordingly, it is a continuous series of connected events in spacetime, where each event is represented by a point on the world line, or the graphical representation of this series.

6. Describe the world line of a light signal.
   Answer: On a standard 2D spacetime diagram (where the temporal coordinate is represented by ct rather than t, the spatial coordinate is represented by x, and the length scale on the spatial and temporal axes is identical) the world line of a light signal is the straight line that makes an angle of $\pm 45°$ with the x axis. Algebraically, using the compact spacetime coordinate notation, this line is represented by the equation: $x^0 = \pm x^1 + C$ where C is a constant.

7. To which concept in the normal "spatial" space the world line concept corresponds?
   Answer: "World line" concept is the counterpart of the "trajectory" concept in normal "spatial" space. So, world line is the "spacetime trajectory" of an object in "spacetime space", like the "spatial trajectory" of an object in "normal space".

8. What is the world line of a free particle in the Minkowski spacetime?
   Answer: A free particle in the Minkowski spacetime will follow a geodesic path. This should be obvious since a free particle will follow a straight path which is a geodesic in this Euclidean flat space. We note that being a straight line in a Euclidean space is a necessary and sufficient condition for being a geodesic.

9. Compare the world line of a free massive particle with the world line of a light ray.
   Answer: The world line of a free massive particle is a geodesic in the Minkowski spacetime manifold, while the world line of a light ray is a null geodesic in the Minkowski spacetime manifold. Being a geodesic (in both cases) is justified by the answer of the previous question, while being a null geodesic (in the second case) can be inferred from the mathematical expression of spacetime interval where the temporal change is equal in magnitude and opposite in sign to the spatial change and hence they cancel each other resulting in a null interval (see § 3.9.3).

10. Compare the concepts of "null geodesic" and "geodesic" in the Minkowski spacetime.
    Answer: In brief, null geodesic is a special case of geodesic in the Minkowski spacetime, because a geodesic can be the trajectory of light (and hence it is null) or the trajectory of a free massive object (and hence it is not null).

11. What is the world line of a massive particle moving under the influence of an external force?
    Answer: Under the influence of an external force, a massive particle will move in the spacetime along a curve. Hence, its world line is not a geodesic in the flat space of Minkowski spacetime. In other terms, the motion under the influence of an external force is accelerated and hence the trajectory of the particle in the spacetime is not straight, i.e. it is not geodesic.

3.9.2 Light Cone

1. Make a simple sketch of a 3D (i.e. one temporal and two spatial) light cone diagram where the causal relations with the event of interest are considered in the labels: past,

3.9.2 Light Cone

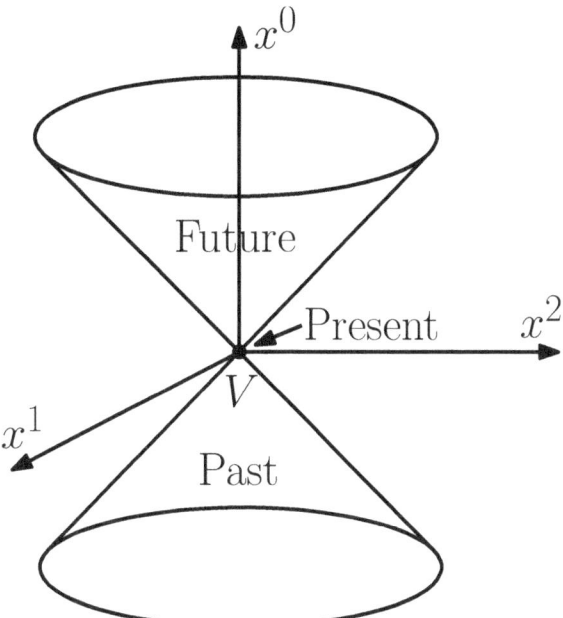

Figure 7: A simple sketch of a 3D light cone diagram of a given event V with one temporal dimension represented by x^0, and two spatial dimensions, represented by x^1 and x^2, where causal relations are considered in the labels past, present and future.

present and future.

Answer: The required sketch is shown in Figure 7. As a matter of terminology about chronological order, all the events in the event space with $x^0 < 0$ belong to the past, all the events in the event space with $x^0 > 0$ belong to the future and all the events in the event space with $x^0 = 0$ belong to the present. However, only the events inside the lower part of the cone[55] can influence V, only the events inside the upper part of the cone can be influenced by V and only the events that are co-positional to V at the present can influence and be influenced by V. Accordingly, only those parts in the event space that can have a causal relation with V are labeled as past, present and future. This may be justified by the claim that "past", "present" and "future" are meaningful only from the perspective of a given event and through causal interactions which may be considered as definition of these terms inline with the philosophical and epistemological framework of special relativity.

2. What are the main characteristics of the light cone diagram and the relations between the events in the event space that are represented by the light cone diagram?

    Answer: The following are some of the characteristics of the light cone diagram and the nature of the relations between the events in its event space:
    • The light cone diagram belongs to a specific event V which is at the apex of the cone and hence it is meaningless to speak about light cone without specifying an event to

---

[55] We note that "inside" in this context should include the surface of the cone if the speed c is viable for establishing communication.

3.9.2 Light Cone

which the light cone belongs.
- The light cone itself (i.e. the surface of the cone not the diagram) is formed of representation of light rays emitted in all spatial directions from the point of origin where the event V is located.[56] This can be caused by an event of a light flash set out in the event space at $\mathbf{x} = (0,0,0,0)$ where V took place with an expanding spherical wave front, where the frame of observation is supposed to be inertial. Accordingly, the light cone is made of null geodesics in the 4D flat space of Lorentz mechanics with the Minkowski spacetime metric (refer to § 6.3).
- The equation of the surface of the light cone in any inertial frame takes the same form, that is:[57]
$$(x^0)^2 - (x^1)^2 - (x^2)^2 - (x^3)^2 = 0$$
where the labels of the spacetime coordinates may be different, e.g. primed and unprimed for primed and unprimed frames.
- All the events inside the two parts of the light cone are timelike with respect to the event of interest V which is at the apex of the cone. All the events outside the two parts of the light cone are spacelike with respect to V. All the events on the surface of the two parts of the light cone are lightlike with respect to V.[58]
- Causal accessibility can occur only by timelike and lightlike trips because in a spacelike trip from V to another event V_1, for example, where V is supposed to cause V_1, the signal will not arrive to the location of V_1 before the occurrence of V_1, since the signal cannot exceed the speed of light c, and hence V cannot cause V_1. This is also the case when V is supposed to be caused by V_1.
- The past, present and future in the event space in general are relative to the time of the event of interest which is represented by the point at the apex of the light cone. However, if causal relations with the event of interest are considered then only those parts of the event space which are inside or on the surface of the light cone can be described as past, present and future. Accordingly, the null cone divides only the part of the spacetime that is contained in the cone or on its surface into three mutually exclusive regions: past, present and future (refer to Figure 7) while the part of the spacetime that is outside the cone is labeled "elsewhere".
- If two events are positioned in one of the three regions (i.e. inside the lower part of the cone, or inside the upper part of the cone, or outside the cone) then they can be connected by a continuous curve that is totally contained inside that region with no common point with the cone. Moreover, if two events are located in the lower/upper part, then they can also be connected by a straight line segment that is completely contained in the lower/upper part with no common point with the cone.
- If two events are located in two different regions of the above three regions (i.e. inside lower part, inside upper part, and outside cone) then any continuous curve that

---

[56] As indicated earlier, the lower part of the cone is just a reflection in the $x^0 = 0$ plane of the upper part which represents the actual propagation of this signal in spacetime.

[57] If the origin of the frame is not at the apex then a differential form (using Δ) is required.

[58] In fact, these labels are usually attached to the trips or intervals that connect V to the points in these regions.

connects these events must cut the cone in one point at least.

We should remark that the above characterization of the region outside the light cone as being representing events that are causally inaccessible to the event of interest, which is at the apex of the light cone, and hence they cannot influence or be influenced by that event is a consequence of the assumption that the characteristic speed of light c is the ultimate speed, and hence no signal that communicates between two events in the spacetime to establish a causal relation between them can travel faster than c. As indicated before, this limitation is subject to the limitations on the domain of Lorentz mechanics and potential limitations on its current formulation and hence it should not be understood as an implicit unconditional acceptance of the proposition that c is a restricted and ultimate physical speed. We should also assume the acceptance of the premise that causal relations are established by communication signals which are subject to the aforementioned speed restrictions.

3. Outline the two main types of light cone diagram with regard to the number of dimensions.

   Answer: Any light cone diagram should have a temporal dimension. Moreover, it should have at least one spatial dimension and no more than two spatial dimensions due to the restriction on the number of dimensions that can be represented on a paper or screen (or indeed even on a 3D model if the dimensions of the model are supposed to independently represent all the four dimensions of the 4D spacetime). Accordingly, we have two main types of light cone diagram: a 2D diagram with one temporal dimension and one spatial dimension (i.e. $x^0 x^1$), and a 3D diagram with one temporal dimension and two spatial dimensions (i.e. $x^0 x^1 x^2$). In fact, the 2D light cone diagram is just a vertical cross section of the 3D light cone diagram created by the plane that contains the x^0 coordinate, i.e. the $x^0 x^1$ plane of the $x^0 x^1 x^2$ space. Moreover, labeling the 2D diagram as light *cone* diagram may not be appropriate although it may be used in the literature.

4. Discuss if the use of light cone diagram is dependent on accepting special relativity and its second postulate.

   Answer: No, it is not dependent although some of its interpretations and implications may depend. In particular, the invariance of light cone (i.e. taking the same conical shape in all inertial frames) is based on the invariance of spacetime interval which is based on the formalism of Lorentz mechanics in general regardless of any particular interpretation such as special relativity. However, any particular interpretation has certain consequences and implications that should be inspected carefully and hence they are accepted or rejected based on the criteria of logical consistency, the principles of reality and truth and experimental evidence. It should be remarked that the consistency and aesthetic factors of any mathematical device, like light cone, cannot be used as evidence to support or dismiss a scientific theory (whether formalism or interpretation) without experimental evidence. These devices may be used as useful tools for demonstrating and conceptualizing a theory or even inferring some logical results from it but they cannot stand as independent evidence for or against any scientific theory. Scientific theories should be accepted or rejected according to real experimental evidence and not according to mathematical artifacts although it is completely legitimate to create and use such mathematical artifacts to serve an established theory.

5. What a cross section of the light cone made by a plane perpendicular to the x^0 axis at a given value of x^0 represents?
Answer: It represents a cross section of the sphere representing the position of the wave front of the light in a 3D spatial space. This cross section is a great circle of the sphere and hence it is circular in shape and is centered at the spatial origin where the apex of the cone. For a 3D light cone diagram (i.e. $x^0 x^1 x^2$ like Figure 7), the plane represents the $x^1 x^2$ plane of the 3D spatial space and hence the circle represents the trace of the wave front on the xy plane at a given instant of time t corresponding to the given value of the x^0 ($= ct$) coordinate.

6. Put appropriate restrictions (according to the Lorentzian speed restrictions) on the world line of (A) a light ray passing through a given point in spacetime and (B) a massive object passing through such a point.
Answer:
(A) The world line of light ray should be on the light cone of that point.
(B) The world line of massive object should be inside the light cone of that point.

3.9.3 Spacetime Interval

1. Define, descriptively and mathematically, the concept of spacetime interval using its infinitesimal form.
Answer: Spacetime interval in its infinitesimal form $d\sigma$ is the infinitesimal "line element" in spacetime and hence it is the equivalent of infinitesimal line element ds in ordinary "spatial" space. Both these represent the metric of their manifold, i.e. spacetime for the first and space for the second. Now, since the line element ds in normal space is defined as:

$$(ds)^2 = (dx^1)^2 + (dx^2)^2 + (dx^3)^2$$

then the spacetime interval $d\sigma$ should be defined as:[59]

$$(d\sigma)^2 = (dx^0)^2 - (dx^1)^2 - (dx^2)^2 - (dx^3)^2$$

The spacetime interval may also be defined as spatial minus temporal, that is:[60]

$$(d\sigma)^2 = (dx^1)^2 + (dx^2)^2 + (dx^3)^2 - (dx^0)^2$$

and this does not change the invariance property of the spacetime interval because the latter definition just alters the sign of the square of the interval across the frames. As we see, the space and time coordinates in the definition of the spacetime interval should

---

[59] The difference in sign between the temporal and spatial coordinates is based on the requirement of the invariance of the spacetime interval which is ultimately based on the use of the characteristic speed of light for calibrating the measurements of the spacetime coordinates. This also reflects the metric of the Minkowski space.
[60] Or rather: $(d\sigma)^2 = (dx^1)^2 + (dx^2)^2 + (dx^3)^2 - (dx^4)^2$.

3.9.3 Spacetime Interval

have opposite signs if the interval should be invariant (see § 3.9.4). This is in accord with the metric of the Minkowski space, as we will see in § 6.3.

2. What differences you note between the space interval ds and the spacetime interval $d\sigma$? Also, what are the common features between ds and $d\sigma$?
   Answer: We note the following differences:
   - While ds belongs to a 3D manifold, $d\sigma$ belongs to a 4D manifold.
   - While all the terms of ds are positive, $d\sigma$ has a mix of positive and negative terms.
   - Based on the previous point, while ds is positive definite,[61] $d\sigma$ is not positive definite. In fact, $d\sigma$ can even be imaginary, i.e. the spacetime interval is positive when its square is positive, zero when its square is zero, and imaginary when its square is negative.

   One of the common features between ds and $d\sigma$ is that they both represent the metric of their spaces. Also, they are both invariant under the relevant transformations of their spaces.

3. Is the spacetime of Lorentz mechanics flat or curved? Explain why.
   Answer: The spacetime of Lorentz mechanics is flat. The reason is that an nD space is flat *iff* it is possible to find a coordinate system for the space in which the line element ds is given by:[62]

   $$(ds)^2 = \zeta_1(dx^1)^2 + \zeta_2(dx^2)^2 + \ldots + \zeta_n(dx^n)^2 = \sum_{i=1}^{n} \zeta_i(dx^i)^2$$

   where the indexed ζ are ± 1 while the indexed x are the coordinates of the space. This condition is obviously satisfied by the spacetime interval $d\sigma$ as given by its equation in the text and in the answer of a previous question. We note that this condition is equivalent to having a diagonal metric tensor with all its diagonal elements being ± 1.

4. Discuss if the invariance of the spacetime interval across all inertial frames under the Lorentz transformations is based on accepting special relativity and its second postulate.
   Answer: The invariance of spacetime interval under the Lorentz spacetime coordinate transformations is based on the formalism of Lorentz mechanics regardless of any particular interpretation such as special relativity. As discussed earlier and will be discussed further, we approach Lorentz mechanics in its pure formalism where the Lorentz spacetime coordinate transformations are the only postulates of this formalism regardless of any particular interpretation or philosophical or epistemological framework. However, as indicated earlier and will be discussed further in the future, the invariance of spacetime interval under the Lorentz transformations is based on using the characteristic speed of light for calibrating the measurement of spacetime coordinates in this mechanics. In fact, the invariance of the speed of light across all inertial frames is also a consequence of this use and hence it is equivalent to the invariance of spacetime interval.

---

[61] Assuming it represents an interval between two distinct points in the space.
[62] We note that ds here is used to symbolize the infinitesimal line element of any space and not necessarily the ordinary 3D space.

3.9.3 Spacetime Interval

5. Write the expression of the lightlike interval in a way that makes it more sensible.
   Answer: If for notational simplicity we use x, y, z for the spatial variations and ct for the temporal variation in the expression of the spacetime interval, then the equation of the lightlike interval will make more sense if we write it in the following form:
   $$t = \frac{\sqrt{x^2 + y^2 + z^2}}{c}$$
   since it states that the time of the interval is equal to the distance traveled by light divided by the speed of light. We can similarly write it as:
   $$c = \frac{\sqrt{x^2 + y^2 + z^2}}{t}$$
   which is a formal statement that the speed of light is equal to the distance traveled divided by the time of travel. Both these forms make full sense within the framework of Lorentz mechanics regardless of any particular interpretation. The invariance of the speed of light may be inferred from this but it is not necessarily in the sense of the second postulate of special relativity. More fundamentally, the above expressions (especially the first) indicate the use of c for calibrating the measurement of spacetime coordinates. More about this and the interpretation of the invariance of the speed of light and if it is real or apparent will be discussed in detail later in the book.

6. List and describe the types of spacetime interval.
   Answer: The spacetime interval between two events, $V_1(\mathbf{x}_1)$ and $V_2(\mathbf{x}_2)$, in the event space is classified as:
   - Spacelike when distance exceeds time, that is:
   $$\left(\Delta x^0\right)^2 < \left(\Delta x^1\right)^2 + \left(\Delta x^2\right)^2 + \left(\Delta x^3\right)^2$$
   - Timelike when time exceeds distance, that is:
   $$\left(\Delta x^0\right)^2 > \left(\Delta x^1\right)^2 + \left(\Delta x^2\right)^2 + \left(\Delta x^3\right)^2$$
   - Lightlike when time and distance are equal, that is:
   $$\left(\Delta x^0\right)^2 = \left(\Delta x^1\right)^2 + \left(\Delta x^2\right)^2 + \left(\Delta x^3\right)^2$$
   The above classification of the spacetime interval between two events (i.e. spacelike, timelike and lightlike) is independent of the frame, and hence the type of the spacetime interval is invariant under Lorentz transformations from one inertial frame to another inertial frame. This invariance of type is based on the invariance of the spacetime interval itself under Lorentz transformations, as will be shown in the future (see § 3.9.4).

7. Compare the terms: timelike, lightlike and spacelike with the terms: past, present, future and elsewhere.
   Answer: The first set of terms are usually used to label intervals in spacetime, while

3.9.3 Spacetime Interval

the second set of terms are used to label regions in spacetime. However, some terms in these two sets may be used interchangeably where they match, e.g. elsewhere region may be called spacelike region since it is the region where the intervals (connecting the vertex of light cone to the points in that region) there are classified as spacelike.

8. Calculate the spacetime intervals between the following pairs of physical events: (a) V_1 and V_2, (b) V_1 and V_3 and (c) V_2 and V_3 where $V_1\,(13,-2,6,11)$, $V_2\,(10,9,2,-1)$ and $V_3\,(0,1,2,-1)$ and the parenthesized numbers represent (x^0, x^1, x^2, x^3). Also, determine the type of each interval.
   Answer:
   (a) V_1 and V_2:
   $$(\Delta\sigma)^2 = (13-10)^2 - (-2-9)^2 - (6-2)^2 - (11+1)^2 = -272$$
   This is a spacelike interval because $(\Delta\sigma)^2 < 0$.
   (b) V_1 and V_3:
   $$(\Delta\sigma)^2 = (13-0)^2 - (-2-1)^2 - (6-2)^2 - (11+1)^2 = 0$$
   This is a lightlike interval because $(\Delta\sigma)^2 = 0$.
   (c) V_2 and V_3:
   $$(\Delta\sigma)^2 = (10-0)^2 - (9-1)^2 - (2-2)^2 - (-1+1)^2 = 36$$
   This is a timelike interval because $(\Delta\sigma)^2 > 0$.

9. What is the type of the spacetime interval between two simultaneous events and between two co-positional events?
   Answer: The spacetime interval between two simultaneous events is spacelike because $\Delta x^0 = 0$ and hence $(\Delta\sigma)^2 < 0$, while the spacetime interval between two co-positional events is timelike because $\Delta x^i \Delta x^i = 0$ ($i = 1, 2, 3$ with summation over i) and hence $(\Delta\sigma)^2 > 0$. We should remark that in this answer we consider the attributes "simultaneous", "co-positional" and "identical" (refer to § 1.3) as mutually exclusive and hence simultaneous should not be co-positional and co-positional should not be simultaneous.

10. Two planets, A and B, are separated by a distance $d = 9 \times 10^{10}$ kilometers. Two teams of scientists are holding two conferences: the conference of the first team is on planet A which starts at 8:00 o'clock and lasts for 60 minutes, and the conference of the second team is on planet B which starts at 9:30 o'clock and lasts for 30 minutes. Is it possible for a member of the first team to fully attend both conferences? What about partial attendance?
    Answer: We need to determine if it is possible to travel from A to B in 30 minutes, i.e. from the end of the first conference at 9:00 to the start of the second conference at 9:30. Now, the time for a light signal to go from A to B is:
    $$t = \frac{d}{c} \simeq \frac{9 \times 10^{10}}{3 \times 10^5} = 3 \times 10^5 \text{ s}$$
    which is about 83.33 minutes. So according to the formalism of Lorentz mechanics, where physical speeds of massive objects are not allowed to reach or exceed c, it is

3.9.3 Spacetime Interval

impossible to fully attend both conferences. However, it is possible to attend the start of the first conference and the entire second conference but this is partial attendance of both conferences. We also need to assume that for the case of partial attendance the concerned member has the transport to move at any required speed lower than c.

In fact, this question in essence is about the type of the spacetime interval between two events: the end of the first conference and the start of the second conference (for full attendance) which is obviously a spacelike interval, or the start of the first conference and the start of the second conference (for partial attendance) which is obviously a timelike interval.

11. Assuming that causal relations are subject to the speed restrictions, correlate the type of spacetime interval to the causal relation that potentially exist between the two events linked by the interval.
   Answer: Causal relation can exist between two events linked by a timelike or a lightlike interval but not by a spacelike interval.

12. Analyze the significance of the sign of the square of spacetime interval $(\Delta\sigma)^2$ between two events and its invariance across inertial frames. Also, try to demonstrate the analysis by using the light cone of an event.
   Answer: Since $(\Delta\sigma)^2$ is invariant, its sign is also invariant and hence in all frames we should have either $(\Delta\sigma)^2 < 0$ or $(\Delta\sigma)^2 = 0$ or $(\Delta\sigma)^2 > 0$. Accordingly:[63]
   • If $(\Delta\sigma)^2 < 0$ then there is no inertial frame in which the two events are identical or co-positional. Moreover, there should be a frame in which the two events are simultaneous.
   • If $(\Delta\sigma)^2 = 0$ then there is no inertial frame in which the two events are simultaneous or co-positional. Now, because being identical is invariant across frames (see § 7.5), then they should be either identical in all frames or anti-identical in all frames. However, if they are anti-identical then they should be communicated by a light signal for any causal relation (assuming speed restrictions), i.e. they are connected by a null geodesic.
   • If $(\Delta\sigma)^2 > 0$ then there is no inertial frame in which the two events are identical or simultaneous. Moreover, there should be a frame in which the two events are co-positional.
   In brief:
   (a) $(\Delta\sigma)^2 < 0$: simultaneous or anti-identical.
   (b) $(\Delta\sigma)^2 = 0$: identical in all frames or anti-identical in all frames.
   (c) $(\Delta\sigma)^2 > 0$: co-positional or anti-identical.
   The above classification can be demonstrated by the light cone $x^0 x^1 x^2$ of an event V (which is at the apex) where we have:
   (A) $(\Delta\sigma)^2 < 0$: this is represented by the region outside the light cone (spacelike) where the events are either simultaneous to V (represented by the $x^1 x^2$ plane excluding the apex) or anti-identical to V (represented by the other parts of that region).
   (B) $(\Delta\sigma)^2 = 0$: this is represented by the cone itself (lightlike) including the apex where the events are either identical to V (represented by the apex) or anti-identical

---

[63] The case of being anti-identical (refer to § 1.3) is not considered here because it is general and applies in all cases. Moreover, we consider these categories (i.e. simultaneous, co-positional, identical and anti-identical) as mutually exclusive.

to V (represented by the other parts of the cone) and this (i.e. being identical or anti-identical) should be true in all frames.

(C) $(\Delta\sigma)^2 > 0$: this is represented by the region inside the light cone (timelike) where the events are either co-positional to V (represented by the x^0 axis excluding the apex) or anti-identical to V (represented by the other parts of that region).

3.9.4 Invariance of Spacetime Interval

1. What is the essential condition for the invariance of spacetime interval between frames of reference?
   Answer: The frames should be inertial. This is a general assumption in all the formulations of Lorentz mechanics as well as classical mechanics. This invariance is obviously with respect to the Lorentz transformations, i.e. the spacetime interval is invariant across *inertial frames* under the *Lorentz transformations* of spacetime coordinates.

2. What it means to say that the infinitesimal line element of a space is invariant? How this applies to the spacetime interval $d\sigma$?
   Answer: It means that the infinitesimal line element of the given space, which is usually symbolized by ds, will have the same length regardless of the employed coordinate system and hence we should have:
   $$ds = ds'$$
   where the primed and unprimed symbols represent the line element in the primed and unprimed coordinate systems.[64] Now, since the spacetime interval is the line element of the spacetime manifold, it means that it will have the same value regardless of the employed reference frame and hence we should have:
   $$d\sigma = d\sigma'$$
   where the primed and unprimed symbols represent the spacetime interval in the primed and unprimed reference frames of two inertial observers. This invariance is obviously under the Lorentz transformations of spacetime coordinates.

3. What is the significance of the invariance of spacetime interval in Lorentz mechanics?
   Answer: The significance is that although two inertial observers may disagree on the spatial separation between two physical events or the size of the time interval between these events (as well as their disagreement on the spacetime coordinates of the events), they will agree on their spacetime separation which is represented by their spacetime interval. This means that according to Lorentz mechanics, the real space of the events is not the ordinary "spatial" space or the time continuum but it is this mix of "spacetime". Hence, space coordinates and time are interwoven and mingled in this spacetime composition. Therefore, time and space interact with each other and

---

[64] For classical mechanics, this invariance is under the Galilean transformations of space coordinates. As discussed in § 1.8 within the context of the invariance of physical laws, any invariance principle requires a domain of validity (e.g. inertial frames) and a set of transformation rules (e.g. the Galilean transformations of space and time).

the variation of one should result in a variation of the other to keep the invariance of spacetime interval. This is in contrast with the view of classical mechanics where space and time are two separate reciprocally-independent entities and hence the variation in space and the variation in time are independent of each other. We note that at the root of this invariance of spacetime interval are the effects of length contraction and time dilation where they vary in such a way that keeps the spacetime interval invariant. This issue is also linked to the invariance of the speed of light across inertial frames where this invariance is ultimately based on the use of the characteristic speed of light for calibrating spacetime measurements. These points will be clarified further in the future (see for example § 9.7.1 and § 11).

4. Use the fact that the spacetime interval is invariant to analyze the movement through space and time.
   Answer: The invariance of spacetime interval in its finite form across frames may be expressed as:[65]
   $$\left(\Delta x^0\right)^2 - \left(\Delta x^1\right)^2 - \left(\Delta x^2\right)^2 - \left(\Delta x^3\right)^2 = 0$$
   where Δ represents changes in time and space across different inertial frames which are in relative motion. This means that when a temporal change occurs, it will be compensated by an opposite spatial change to keep the total change zero and vice versa. So, moving at a higher rate in time means moving at a lower rate in space and vice versa.[66] Accordingly, standing still in space (i.e. zero spatial change) means maximum time change. Similarly, moving at the maximum rate in space (which is c according to the current formulation of Lorentz mechanics) means minimum time change. This may be linked to the length contraction and time dilation effects (as seen from an external inertial frame) where the first represents spatial change while the second represents temporal change. These changes originate from opposite physical effects since in the first the length contracts (or decreases) while in the second the time dilates (or increases). They can also be seen in the opposite sense where the length dilates (or expands) while the time contracts (or shrinks). As we will see later in the book, this is based on a difference in the proper-improper perspective and hence the two effects will be essentially the same (i.e. both contraction or both dilation) when we unify this perspective. These issues will be discussed in detail in the future.

5. Show that ds and dt are separately invariant under the Galilean transformations.
   Answer: It was shown in the exercises of § 1.9.1 that the distance in the space of classical mechanics is invariant under the Galilean transformations. However, we demonstrate it here again using a more general form although we still use standard setting to be able to use the Galilean transformations in the form given in the text noting that this

---
[65] We are not using lightlike interval although the equation may suggest this.
[66] In other words, when the difference in time/space between two frames increases, the difference in space/time between these frames decreases.

3.9.4 Invariance of Spacetime Interval

does not affect the generality of the argument:[67]

$$\begin{aligned}(ds')^2 &= (dx')^2 + (dy')^2 + (dz')^2 \\ &= (x'_2 - x'_1)^2 + (y'_2 - y'_1)^2 + (z'_2 - z'_1)^2 \\ &= ([x_2 - vt] - [x_1 - vt])^2 + (y_2 - y_1)^2 + (z_2 - z_1)^2 \\ &= (x_2 - x_1)^2 + (y_2 - y_1)^2 + (z_2 - z_1)^2 \\ &= (dx)^2 + (dy)^2 + (dz)^2 \\ &= (ds)^2\end{aligned}$$

In fact, this should be obvious without this formal demonstration because distance is invariant in the 3D Euclidean space of classical mechanics where the Galilean transformations are based on such a space. Similarly, for dt:

$$dt' = dt$$

where the Galilean time transformation (i.e. $t' = t$), which is based on the universality of time, is used (refer to § 1.9.1).

Therefore, we conclude that ds and dt are separately invariant under the Galilean transformations. This, added to the previous demonstration that the spacetime interval is not invariant under the Galilean transformations, means that space and time are two independent manifolds that underlie classical mechanics.

6. Show that ds and dt are not invariant under Lorentz transformations.

   Answer: This should be obvious due to length contraction and time dilation effects in Lorentz mechanics, as we will see in the future. However, we show this formally as follows where we use the upcoming Lorentz spacetime coordinate transformations which are based on a state of standard setting (refer to § 4.2.1):

$$\begin{aligned}(ds')^2 &= (dx')^2 + (dy')^2 + (dz')^2 \\ &= (x'_2 - x'_1)^2 + (y'_2 - y'_1)^2 + (z'_2 - z'_1)^2 \\ &= (\gamma [x_2 - vt] - \gamma [x_1 - vt])^2 + (y_2 - y_1)^2 + (z_2 - z_1)^2 \\ &= \gamma^2 (x_2 - x_1)^2 + (y_2 - y_1)^2 + (z_2 - z_1)^2 \\ &= \gamma^2 (dx)^2 + (dy)^2 + (dz)^2 \\ &\neq (ds)^2 = (dx)^2 + (dy)^2 + (dz)^2\end{aligned}$$

Also:

$$\begin{aligned}dt' &= t'_2 - t'_1 \\ &= \gamma \left(t_2 - \frac{vx}{c^2}\right) - \gamma \left(t_1 - \frac{vx}{c^2}\right) \\ &= \gamma (t_2 - t_1) \\ &\neq dt = t_2 - t_1\end{aligned}$$

---

[67] We note that t is the same for the transformation of x'_1 and x'_2 in the following equations because the spatial coordinates in the measurement of spatial separation between two locations in space should be measured at the same time.

This shows that ds and dt are identical in the two frames only if $\gamma = 1$, i.e. the two frames are in a state of relative rest and hence they are essentially identical. This, added to the previous demonstration that the spacetime interval is invariant under the Lorentz transformations, means that, unlike classical mechanics where space and time are two independent manifolds that underlie this mechanics, in Lorentz mechanics space and time form parts of this merged composition of spacetime manifold whose interval is invariant while the intervals of its spatial and temporal parts are not. The reader should notice the difference by the γ factor in the above expressions which originates from length contraction in the first and from time dilation in the second. We should also note that in the transformation of the x' coordinates of the spatial interval ds' the time t is fixed while in the transformation of the t' coordinates of the temporal interval dt' the spatial coordinate x is fixed. This should be obvious since spatial measurements should take place at the same time while temporal measurements should take place in the same location.

7. Compare and discuss in general terms the issue of invariance in classical mechanics and in Lorentz mechanics.
   Answer: We outline this issue within some examples and cases as follows:
   • The invariance property of classical mechanics is with respect to the Galilean transformations while the invariance property of Lorentz mechanics is with respect to the Lorentz transformations. However, the domain of the invariance property, which is inertial frames, is the same in both mechanics.
   • While the space interval ds and the time interval dt are invariant separately in classical mechanics, the spacetime interval $d\sigma$ is invariant in Lorentz mechanics. On the other hand, $d\sigma$ is not invariant in classical mechanics, and ds and dt are not separately invariant in Lorentz mechanics.
   • In both mechanics we have form invariance of physical laws and value invariance of physical quantities. However, the two mechanics agree about these invariances in some cases but not in all cases. For example, the two mechanics agree on the form invariance of the principle of conservation of momentum or Newton's second law[68] but they do not agree on the invariance of electromagnetic wave equation since this equation is form invariant only in Lorentz mechanics (refer to § 12.3.3). Similarly, the two mechanics agree on the value invariance of mass (considering the modern convention in Lorentz mechanics where mass is an inherent frame-independent quantity or considering rest mass) but they disagree on the value invariance of time interval or length which are invariant only in classical mechanics since in Lorentz mechanics they are subject to time dilation and length contraction effects. They also disagree on the value invariance of spacetime interval which is invariant only in Lorentz mechanics.

---

[68] We note that certain amendments to the classical forms should be imposed for this invariance in Lorentz mechanics to apply.

Chapter 4
Formalism of Lorentz Mechanics

4.1 Physical Quantities

4.1.1 Length

1. What is the physical origin of the relation between the proper and improper values of the length of a physical object according to Lorentz mechanics? What is the restriction on this relation?
   Answer: It is length contraction effect. This effect takes place only in the direction of motion and hence the dimensions of the object in the perpendicular directions to the motion will keep their proper values.

2. A moving object is observed to be contracted in the direction of motion by 40%. Find its relative speed.
   Answer: We should have $\frac{L}{L_0} = 1 - 0.4 = 0.6$, that is:
   $$L_0 = \gamma L$$
   $$\frac{1}{\gamma} = \frac{L}{L_0} = 0.6$$
   $$\sqrt{1-\beta^2} = 0.6$$
   $$1 - \beta^2 = 0.36$$
   $$\beta^2 = 0.64$$
   $$\beta = 0.8$$
   $$v = 0.8c$$

3. A square whose proper area is 1 is seen by an inertial observer to have an area of 0.85. Assuming that the square is moving uniformly with two of its sides oriented along the direction of motion, what is its speed relative to the observer? Repeat the question assuming this time that one diagonal of the square is oriented along the direction of motion.
   Answer: If the proper area of the square is 1 then the proper length of its sides is 1. Now, since the length of the sides which are perpendicular to the direction of motion are not affected by length contraction then the sides which are in the direction of motion should have contracted to 0.85 of their proper length, that is:
   $$L_0 = \gamma L$$
   $$\frac{1}{\gamma} = \frac{L}{L_0} = 0.85$$
   $$\sqrt{1-\beta^2} = 0.85$$
   $$1 - \beta^2 = 0.7225$$

$$\beta^2 = 0.2775$$
$$\beta = \sqrt{0.2775}$$
$$v = c\sqrt{0.2775}$$

Regarding the second part of the question, the proper length of the diagonal of the square is $\sqrt{2}$. Now, the length of the diagonal which is perpendicular to the direction of motion, d_1, is not affected by the motion while the diagonal which is in the direction of motion, d_2, is contracted by a certain factor to make the area shrinks to 0.85. So, instead of a square we now have a rhombus. As it is known, the area of a rhombus is half the product of its diagonals, that is:

$$0.85 = \frac{d_1 d_2}{2}$$
$$d_2 = \frac{2 \times 0.85}{d_1}$$
$$d_2 = \frac{2 \times 0.85}{\sqrt{2}}$$
$$d_2 = 0.85\sqrt{2}$$
$$\frac{d_2}{d_1} = 0.85$$

Now, d_1 is equal to the proper length of the contracted diagonal d_2. Therefore, we can use the first part of the question to conclude that the speed is also $v = c\sqrt{0.2775}$. In fact, this can be concluded from a simple argument that is the square of area 1 in the diagonal orientation is half a square whose sides are of length $\sqrt{2}$ and two of its sides are oriented along the direction of motion, and hence if we assume that the contraction is uniform and applies equally to each infinitesimal piece of the big square then the speed factor should be the same as in the first part of the question. In reality, this is based on the fact that since only one dimension of the moving object is affected by length contraction then the area, like the length, will also be scaled down (or contracted) by the γ factor.

4. O and O' are two inertial frames in a state of standard setting. A stick which is in the xy plane of frame O makes an angle with the x axis of $\pi/4$ in frame O and an angle of $\pi/3$ in frame O'. (a) What is the relative speed v between the two frames? (b) If the proper length of the stick is 1 in frame O what is its length in frame O'?
Answer:
(a) The motion is along the x direction and hence the y direction is not subject to length contraction. Therefore, we should have:

$$\frac{Y}{X} = \tan\frac{\pi}{4} \quad \text{and} \quad \frac{Y'}{X'} = \frac{Y}{X'} = \tan\frac{\pi}{3}$$

where X, Y, X', Y' represent the projection of the stick length on the x and y axes in the unprimed and primed frames. Hence, we have:

$$\frac{(Y/X)}{(Y'/X')} = \frac{(Y/X)}{(Y/X')} = \frac{X'}{X} = \frac{\tan(\pi/4)}{\tan(\pi/3)} = \frac{1}{\sqrt{3}} = \frac{1}{\gamma}$$

4.1.1 Length

that is:

$$\sqrt{1-\beta^2} = 1/\sqrt{3}$$
$$1-\beta^2 = 1/3$$
$$\beta^2 = 2/3$$
$$\beta = \sqrt{2/3}$$
$$v = c\sqrt{2/3}$$

(b) We have:

$$\begin{aligned} L'^2 &= X'^2 + Y'^2 \\ &= \left[\frac{Y'}{\tan(\pi/3)}\right]^2 + Y'^2 \\ &= \left[\frac{Y}{\tan(\pi/3)}\right]^2 + Y^2 \\ &= \left[\frac{\sin(\pi/4)}{\tan(\pi/3)}\right]^2 + \sin^2(\pi/4) \\ &= \sin^2(\pi/4)\left(\frac{1}{3}+1\right) \\ &= \frac{1}{2}\left(\frac{4}{3}\right) \\ &= \frac{2}{3} \end{aligned}$$

and hence:

$$L' = \sqrt{\frac{2}{3}} \simeq 0.8165\,\text{m}$$

5. A cube of sides $s = 1$ m in its rest frame is seen by an inertial observer O to be moving along one of its sides with speed $v = 0.6c$. What are the dimensions of this cube as measured by O? What is its volume?
Answer: According to the length contraction formula, the length of each one of the four sides of the cube which are in the direction of motion will be measured as:

$$L = \frac{s}{\gamma} = \sqrt{1-\frac{v^2}{c^2}} = \sqrt{1-0.36} = 0.8\,\text{m}$$

The sides of the cube in the other two directions (i.e. in the directions which are perpendicular to the motion) will not be affected. Hence, the length of four of the cube sides will be measured as 0.8 m, while the length of the remaining eight sides will remain 1 m as in their rest frame.
Accordingly, the volume of the cube should be:

$$V = 1 \times 1 \times 0.8 = 0.8\,\text{m}^3$$

4.1.1 Length

This can also be easily inferred from the fact that since only one dimension of the moving object is affected by length contraction then the volume, like the length, will also be scaled down (or contracted) by the γ factor.

6. Repeat the previous question but this time the cube is seen to be moving along one of its face diagonals.
   Answer: The cube is schematically depicted in Figure 8 where it is supposed to move in the direction of its face diagonal AC (or ac). It should be obvious that its four sides Aa, Bb, Cc and Dd will not be affected because they are perpendicular to the direction of motion and hence they keep their length as in their rest frame, i.e. $s_1 = 1$. Now, the length of the face diagonals which are perpendicular to the direction of motion (i.e. BD and bd) will not be affected by the motion and hence it is $\sqrt{2}$, while the face diagonals which are in the direction of motion will be shortened, according to the previous exercise, by a factor of 0.8 and hence they will be measured as $0.8\sqrt{2}$. So, the length of each one of the remaining eight sides will be:

$$s_2 = \sqrt{\left(\frac{\sqrt{2}}{2}\right)^2 + \left(\frac{0.8\sqrt{2}}{2}\right)^2} = \sqrt{0.5 + 0.32} \simeq 0.9055\,\text{m}$$

Regarding the volume, it is 0.8 as in the previous question. This can be obtained from calculation by using the dimensions of the contracted cube, that is:

$$V = 1 \times \frac{\sqrt{2} \times 0.8\sqrt{2}}{2} = 0.8\,\text{m}^3$$

It can also be obtained from the simple reasoning that since only one dimension of the cube is affected by length contraction then the volume, like the length, will also be scaled down (or contracted) by the γ factor.

7. The radius of a circle is measured in its rest frame to be 1. The area of this circle is measured to be 2 by a moving observer. What is the speed of the moving observer relative to the rest frame of the circle?
   Answer: The moving observer will see the circle as ellipse with a semi-major radius $a = 1$ and a semi-minor radius b. From the formula of the area A of ellipse we have:

$$\begin{aligned} A &= \pi a b \\ b &= \frac{A}{\pi a} = \frac{2}{\pi} \end{aligned}$$

Now, from the length contraction formula, applied to the radius in the direction of motion, we have:

$$\begin{aligned} L_0 &= \gamma L \\ a &= \gamma b \\ \frac{1}{\gamma} &= \frac{b}{a} \end{aligned}$$

4.1.2 Time Interval

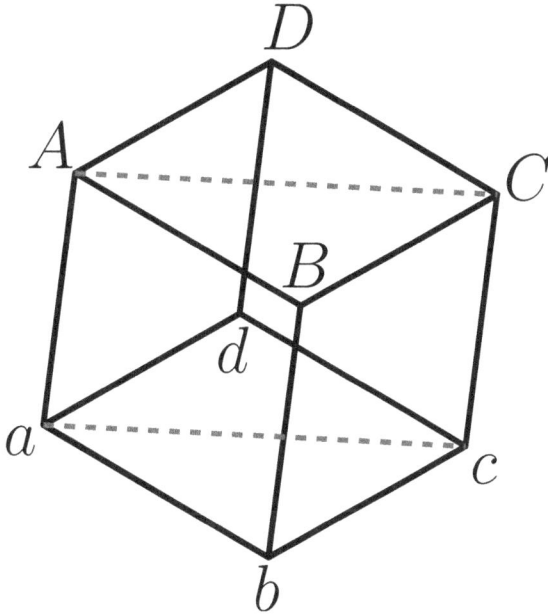

Figure 8: A cube in relative uniform motion along its face diagonal AC (or ac).

$$\sqrt{1-\beta^2} = \frac{2}{\pi}$$
$$\beta^2 = 1 - \frac{4}{\pi^2}$$
$$\beta = \sqrt{1 - \frac{4}{\pi^2}}$$
$$v \simeq 0.7712c$$

4.1.2 Time Interval

1. What is the physical origin of the relation between the proper and improper values of the time interval between two events according to Lorentz mechanics?
   Answer: It is the time dilation effect. According to this effect, the time interval shrinks (like length in the length contraction effect) with the movement. This effect is seen by an external observer as dilation of time, as will be clarified later.
2. Describe in a few words the relation between the proper and improper values of length and time interval according to Lorentz mechanics.
   Answer: The proper length is greater than the improper length due to length contraction, while the proper time interval is shorter than the improper time interval due to time dilation. This can be easily seen from the formulae of length contraction and time dilation (i.e. $L_0 = \gamma L$ and $\Delta t = \gamma \Delta t_0$) since the Lorentz factor γ is greater than 1 when $v \neq 0$. As indicated before and will be clarified further, the proper-improper perspectives in these effects are different and hence they are labeled in this way.
3. Inertial observers O and O' are in a state of relative motion with speed $v = 0.8c$. The

4.1.2 Time Interval

time interval between two events which are co-positional in the frame of O' is measured by O' to be $\Delta t' = 3$. What is the time interval Δt as measured by O?

Answer: Using the time dilation formula, we have:

$$\Delta t = \gamma \Delta t' = \frac{3}{\sqrt{1 - 0.8^2}} = \frac{3}{0.6} = 5$$

4. A cosmonaut is planning to travel at a constant velocity to a planet that is 1 Ly (i.e. light year) away from the Earth in 5 years of his time (i.e. he ages five years during this journey). What is the required speed for this journey? Solve this problem once from the frame of the Earth and once from the frame of the cosmonaut to check if the two solutions are consistent and comment on the results. Assume in your answer that the Earth frame is inertial and it is in a state of rest relative to the distant planet. Also assume that the time dilation effect takes place in the frame of the cosmonaut.

Answer: If we symbolize the distance with d and the time interval with Δt, then the speed (which is the same in both frames since it is the speed of each frame relative to the other frame) can be expressed as:

$$v = \frac{d_E}{\Delta t_E} \qquad \text{or} \qquad v = \frac{d_c}{\Delta t_c}$$

where the first relation represents the speed as seen from the frame of the Earth, while the second relation represents the speed as seen from the frame of the cosmonaut. Now, we are given $d_E = 1$ Ly and $\Delta t_c = 5$ years.

From the frame of the Earth we use time dilation formula to obtain Δt_E which is unknown and hence we have:

$$v = \frac{d_E}{\Delta t_E} = \frac{d_E}{\gamma \Delta t_c} = \frac{1 \times c}{5\gamma}$$

where c (and hence v) is in units of length (say meters) per year and 1 stands for the time of one year. Hence:

$$\frac{1}{\gamma} = 5\beta$$
$$\sqrt{1 - \beta^2} = 5\beta$$
$$1 - \beta^2 = 25\beta^2$$
$$26\beta^2 = 1$$
$$\beta = \sqrt{1/26}$$
$$v \simeq 0.1961c$$

From the frame of the cosmonaut we use length contraction formula to obtain d_c which is unknown and hence we have:

$$v = \frac{d_c}{\Delta t_c} = \frac{(d_E/\gamma)}{\Delta t_c} = \frac{d_E}{\gamma \Delta t_c} \simeq 0.1961c$$

4.1.2 Time Interval

as before.

Comment: time dilation and length contraction produce effects on spacetime coordinates that are equivalent in essence. As well as supporting the conclusions that have been drawn from the previous exercises, this exercise should also shed light on the issue of the invariance of the spacetime interval in Lorentz mechanics which is linked to the invariance of the speed of light and its use in spacetime calibration. It also highlights the issue of the profound merge of space and time in the spacetime manifold where the dimensions of this manifold are subject to similar (or even identical) effects, i.e. contraction or dilation of spacetime, under the influence of motion. These issues will be investigated further in the future.

5. An astronaut fired a rocket from the back of his inertial spaceship in the forward direction. In the frame of the astronaut the length of the spaceship is 100 m and the speed of the rocket is $0.5c$. What is the time required by the rocket to reach the front of the spaceship (a) according to the frame of the astronaut and (b) according to an inertial frame O in which the spaceship is moving along the direction of the rocket with a speed of $0.7c$? Comment on this question.

Answer: We label the frame of the spaceship with O' and assume that O and O' are in a state of standard setting where all the motions are taking place along the common x-x' direction. We also symbolize the speed of the rocket relative to the spaceship with u and the speed of O' relative to O with v, and we use the subscripts b and f to refer to the back and front of the spaceship.

(a) From frame O' we have:

$$\Delta t' = t'_f - t'_b = \frac{x'_f - x'_b}{u} = \frac{100}{0.5c} \simeq 6.6667 \times 10^{-7} \, \text{s}$$

(b) From frame O we should use the basic Lorenz transformations of spacetime coordinates (refer to § 4.2.1) and hence we have:

$$\begin{aligned}
\Delta t &= t_f - t_b \\
&= \gamma\left(t'_f + \frac{vx'_f}{c^2}\right) - \gamma\left(t'_b + \frac{vx'_b}{c^2}\right) \\
&= \frac{(t'_f - t'_b) + \frac{v}{c^2}(x'_f - x'_b)}{\sqrt{1 - (v/c)^2}} \\
&\simeq \frac{6.6667 \times 10^{-7} + \frac{0.7}{c} \times 100}{\sqrt{1 - 0.7^2}} \\
&\simeq 1.2603 \times 10^{-6} \, \text{s}
\end{aligned}$$

Comment: if we compare $\Delta t'$ and Δt we find that they do not satisfy the time dilation formula (i.e. $\Delta t \neq \gamma \Delta t'$). This exercise highlights the fact that the time dilation formula applies when the two events are in the same location in the transformed frame, i.e. they are co-positional. The reason is that if they are not co-positional then there is a change in both time and space and hence the time dilation formula, which is based

4.1.2 Time Interval 112

on having a change in time only, is not suitable to tackle the problem. In this exercise, the two events are at different locations since one is at the back and the other is at the front. Hence, this type of problems should not be solved by the time dilation formula but by the basic Lorentz spacetime coordinate transformations which take into account both the spatial and temporal changes as done above. In brief, time dilation formula applies to co-positional events (i.e. assuming change in time only in the transformed frame) while length contraction formula applies to simultaneous events (i.e. assuming change in space only in the transformed frame).

6. Two inertial observers are in a state of relative motion. The time interval between events as measured by one of these observers scales by a factor of 0.8 by the other observer. What is the relative speed between these observers?
   Answer: We should note first that "The time interval between events...etc." in the question is a hint that this is a relation between the proper and improper values because if this is not the case then the scaling will not be the same due to the involvement of the spatial separation which depends on the spatial coordinates of the two events and hence it will not be the same for all pairs of events that are considered in this question (refer to the previous question). Now, the proper and improper time intervals scale between two frames that are in relative motion either by γ (i.e. $\Delta t = \gamma \Delta t_0$) or by its reciprocal (i.e. $\Delta t_0 = \Delta t / \gamma$). Since $\gamma \geq 1$, 0.8 should be the reciprocal of γ, that is:

$$
\begin{aligned}
0.8 &= \sqrt{1 - \frac{v^2}{c^2}} \\
0.64 &= 1 - \frac{v^2}{c^2} \\
\frac{v^2}{c^2} &= 1 - 0.64 \\
v^2 &= 0.36 c^2 \\
v &= 0.6c
\end{aligned}
$$

7. The lifetime of an elementary particle in its rest frame is Δt. In frame O, which is moving at $v_1 = 0.4c$ relative to the particle rest frame, its lifetime is measured to be 3 μs. What is its lifetime in a frame O' which is moving at $v_2 = 0.5c$ relative to the particle rest frame?
   Answer: Δt is the proper lifetime since it is the time of the rest frame of the particle. Hence, we have:[69]

$$\Delta t = \frac{3}{\gamma_1} = 3\sqrt{1 - 0.4^2} \simeq 2.7495 \,\mu\text{s}$$

---

[69] We note that in exercises like this we seemingly follow (for the purpose of diversity and to simplify the presentation) the general approach in the literature of Lorentz mechanics which is generally based on the special relativistic framework and hence it may not be consistent with other interpretations (see for example § 11). As explained earlier (see § 1.5), in the presentation of formalism we are not assuming or embracing or suggesting any particular interpretation, although the language or the tone may suggest a particular interpretation, and hence the interpretative aspects should be obtained from those parts of the book that are dedicated to the interpretation or explicit about it.

Hence, its lifetime as measured in O' is:

$$\Delta t' = \gamma_2 \Delta t = \frac{3\sqrt{1 - 0.4^2}}{\sqrt{1 - 0.5^2}} \simeq 3.1749\,\mu s$$

8. Discuss the claim that time dilation effect extends to include even the "biological clock" in the body of living organisms and hence traveling creatures age less.
   Answer: First, there is no reliable evidence of such an extension to include biological clocks and hence this claim is unfounded. Time dilation effect which is concluded and derived from very simple physical systems does not extend automatically to the highly complex biological systems. This should be linked to the issue that we discussed previously (see § 3.4) that even if the time rate of fundamental physical processes follows the rate of time flow, this does not necessarily apply to synthetic and composite processes whose clocks may not follow the time flow pattern of the fundamental physical processes.[70] Moreover, even if the time of fundamental physical processes and the time of biological processes are correlated they are not necessarily the same.[71] Hence, any claim of "travelers age less" should be based on independent experimental and observational evidence not by just extending the physical effect of time dilation to biological systems. In fact, even the possibility of "travelers age more" can be true because the high speed motion can affect the biological tissue in such a way that makes it age faster. In other words, there is a possibility that even if the physical effect of time dilation extends automatically to biological systems, it may still be untrue that "travelers age less" because there may be other biological factors that act in the opposite direction and hence counteract this "ageing less" physical effect and may even be more drastic that an overall (i.e. physical plus biological) "ageing more" effect will take place. So, in simple words biological and physical times are affected by different factors and that is why we see different species age at different rates although all the species on the Earth share the same physical time because they share the same frame. This obviously indicates that the biological aging of living tissue does not follow the physical aging which is common to all. In fact, even different individuals of the same species age at different rates especially in the extreme pathological cases where this difference is easily observable. So, the stories about a twin who traveled to a remote galaxy and returned to find himself younger than his stay-at-home brother (which are told like science facts) are no more than science fiction. At least, they need a solid proof to be accepted and classified as science facts.[72]

4.1.3 Mass

1. Discuss the issue of mass as an intrinsic or extrinsic property according to classical

---

[70] We should also refer to another issue that we discussed previously in § 3.4, i.e. the relation between the time rate of the physical process on which the clock is based and the process of the defining concept.
[71] In fact, we may even claim that the profound meaning of "time" in "physical time" and "biological time" is different.
[72] We mean by science fiction and science fact from the perspective of ageing less; otherwise these stories are obviously science fiction.

mechanics and according to Lorentz mechanics.
Answer: According to classical mechanics mass in general is an intrinsic property. According to Lorentz mechanics, the rest mass is also an intrinsic property. However, the non-rest mass according to the old formalism in defining the mass as $m = \gamma m_0$ is an extrinsic property but according to the new formalism (in which the mass of a massive object does not depend on its state of rest or motion) it is an intrinsic property as in classical mechanics.

2. What is the non-rest mass of an electron whose speed is $u = 0.4c$ according to classical mechanics and according to Lorentz mechanics?
Answer: According to classical mechanics and the modern formalism of Lorentz mechanics, the non-rest mass is an intrinsic property and hence it is the same as its proper value, i.e. $m_e \simeq 9.1094 \times 10^{-31}$ kg. However, according to the old formalism of Lorentz mechanics, the non-rest mass is an extrinsic property which is given by:

$$m = \gamma m_0 = \gamma m_e \simeq \frac{9.1094 \times 10^{-31}}{\sqrt{1 - 0.4^2}} \simeq 9.9392 \times 10^{-31} \text{ kg}$$

3. Let assume that we follow the old formalism of Lorentz mechanics about the variability of mass with speed. A massive object is observed to have an improper mass m_1 at speed $u_1 = 0.05c$ and to have an improper mass $m_2 = 1.25 m_1$ at speed u_2. What is u_2?
Answer: We have:

$$\begin{aligned}
m_2 &= 1.25 m_1 \\
\gamma_2 m_0 &= 1.25 \gamma_1 m_0 \\
\gamma_2 &= 1.25 \gamma_1 \\
\sqrt{1 - \beta_2^2} &= 0.8 \sqrt{1 - \beta_1^2} \\
1 - \beta_2^2 &= 0.64 \left(1 - \beta_1^2\right) \\
\beta_2^2 &= 1 - 0.64 + 0.64 \beta_1^2 \\
\beta_2^2 &= 1 - 0.64 + \left(0.64 \times 0.05^2\right) \\
\beta_2 &\simeq 0.6013 \\
u_2 &\simeq 0.6013c
\end{aligned}$$

4.1.4 Velocity

1. What is the definition of velocity in classical mechanics and in Lorentz mechanics? What is the reason for the difference between the two mechanics in the formulation of velocity?
Answer: The basic definition of velocity, as the rate of change of position with respect to time (i.e. $\mathbf{u} = \frac{d\mathbf{r}}{dt}$), is the same in classical mechanics and in Lorentz mechanics. The difference in the mathematical formulation of velocity in the two mechanics is caused by the difference in the rules of spacetime coordinate transformations where classical

4.1.4 Velocity

mechanics is based on the Galilean transformations while Lorentz mechanics is based on the Lorentz transformations. This causes differences in the velocity transformation rules in the two mechanics when the physical setting involves more than one observer and hence the velocity is measured from more than one frame. Accordingly, the quantitative value of velocity in the two mechanics will differ. In brief, although the basic definition of velocity is the same in both mechanics, the transformation rules are different and hence the values are generally different.

2. Given the fact that in classical mechanics the 1D linear momentum of an object of mass m and velocity u is given by $p = mu$ while in Lorentz mechanics it is given by $p = \gamma mu$, what is the velocity of a massive object whose mass is $m = 1$ and whose momentum is $p = 10^6$ according to these mechanics? What is the percentage difference between the velocities in these mechanics? Comment on the results.

Answer:
In classical mechanics we have:
$$u = \frac{p}{m} = \frac{10^6}{1} = 10^6 \, \text{m/s}$$

In Lorentz mechanics we have:
$$\begin{aligned} p &= \gamma mu \\ p &= \frac{mu}{\sqrt{1-(u^2/c^2)}} \\ 10^6 &= \frac{u}{\sqrt{1-(u^2/c^2)}} \\ 10^{12}\left[1-\left(u^2/c^2\right)\right] &= u^2 \\ u^2 + 10^{12}\left(u^2/c^2\right) &= 10^{12} \\ u^2\left[1+\left(10^{12}/c^2\right)\right] &= 10^{12} \\ u^2 &= \frac{10^{12}}{1+(10^{12}/c^2)} \\ u &= \frac{10^6}{\sqrt{1+(10^{12}/c^2)}} \simeq 999994.44 \, \text{m/s} \end{aligned}$$

where only the positive root is taken because p is given as positive.
The relative difference between the classical and Lorentzian velocities, u_c and u_l, is:
$$\Delta u_r = \frac{u_c - u_l}{u_c} \simeq \frac{10^6 - 999994.44}{10^6} \simeq 0.0000056$$

and hence the percentage difference between the velocities in these mechanics is about 0.00056% which is very tiny.

Comment: even at this very high velocity according to the classical standards, the difference is extremely small. This should be appreciated if we note that this velocity (whether 10^6 or 999994.44) is a very tiny fraction of the speed of light (about 0.33% of c) and hence $\beta \simeq 0.0033$ which is very close to zero and $\gamma \simeq 1.0000056$ which is very close to unity.

4.1.5 Acceleration

1. Discuss the concept and mathematical formulation of acceleration in classical mechanics and in Lorentz mechanics.
   Answer: The basic definition of acceleration, as the rate of change of velocity with respect to time, is the same in both mechanics, that is:

   $$\mathbf{a} = \frac{d\mathbf{u}}{dt} = \frac{d^2\mathbf{r}}{dt^2}$$

   However, there is a difference in the mathematical formulation of acceleration as seen by different observers in the two mechanics. This difference is caused by the difference in the rules of spacetime coordinate transformations where classical mechanics is based on the Galilean transformations while Lorentz mechanics is based on the Lorentz transformations. Accordingly, while in classical mechanics the acceleration of an object is invariant and hence it is the same for all inertial observers (refer to § 1.9.4), in Lorentz mechanics it is not so and hence it is frame dependent (refer to § 4.2.4). However, we will see in § 4.2.4 that this does not affect the inertiality status of frames in classical and Lorentz mechanics and hence any frame should be inertial in both mechanics or non-inertial in both mechanics because even in Lorentz mechanics if the acceleration vanished in one frame it will vanish in all other inertial frames (and hence if it did not it will not).

4.1.6 Momentum

1. Find the momentum of a proton whose velocity is $u = 0.35c$ according to classical mechanics and according to Lorentz mechanics.
   Answer:
   Classical momentum:

   $$p_c = m_p u \simeq 1.6726 \times 10^{-27} \times 0.35 \times 3 \times 10^8 \simeq 1.7563 \times 10^{-19}\,\mathrm{kg.m/s}$$

   Lorentzian momentum:

   $$p_l = \gamma m_p u \simeq \frac{1.6726 \times 10^{-27} \times 0.35 \times 3 \times 10^8}{\sqrt{1 - 0.35^2}} \simeq 1.8748 \times 10^{-19}\,\mathrm{kg.m/s}$$

2. What is the mass of an object whose momentum is $p = 10^8$ and whose velocity is $u = 10^7$ according to classical mechanics and according to Lorentz mechanics? Also, find the percentage error in the classical value.
   Answer:
   According to classical mechanics:

   $$m_c = \frac{p}{u} = \frac{10^8}{10^7} = 10\,\mathrm{kg}$$

According to Lorentz mechanics:

$$m_l = \frac{p}{\gamma u} = \frac{10^8}{10^7}\sqrt{1-(10^{14}/c^2)} \simeq 9.9944\,\text{kg}$$

Relative error:

$$\Delta m_r = \frac{m_c - m_l}{m_c} = \frac{(p/u) - (p/\gamma u)}{p/u} = 1 - \frac{1}{\gamma} = 1 - \sqrt{1-(10^{14}/c^2)} \simeq 0.00056$$

Hence, the percentage error is about 0.056%.

3. Given that the mass of neutron is $m_n \simeq 939.57\,\text{MeV}/c^2$, find the Lorentzian momentum of a neutron whose velocity is $u = 0.55c$ in units of MeV/c.
   Answer: We have:

$$\begin{aligned}
p &= \gamma m_n u \\
&= \gamma\left(939.57\,\text{MeV}/c^2\right) u \\
&= \gamma\left(939.57\,\text{MeV}/c\right)(u/c) \\
&= \gamma\left(939.57\,\text{MeV}/c\right)\beta \\
&= \frac{(939.57\,\text{MeV}/c) \times 0.55}{\sqrt{1 - 0.55^2}} \\
&\simeq 618.76\,\text{MeV}/c
\end{aligned}$$

4.1.7 Force

1. What is the assumption about the mass in the following Lorentzian form of Newton's second law:

$$f = ma\gamma^3$$

   Answer: The assumption is that the mass of the physical object is constant, i.e. the mass does not vary due to exchange between the object and its environment that results in loss or gain such as by ejection of gases from a rocket or attachment of snow to a rolling snow ball. We remark that in the above equation we follow the modern formalism of Lorentz mechanics about mass where it is considered an invariant (i.e. frame independent) property. This can be judged from the expression of momentum from which this equation is obtained, as shown in the text.

2. What is the force that acts on an object whose mass is $m = 1$ and whose trajectory along the x axis is described by the following equation:

$$x = 3t + 12$$

   where t is time? Comment on your answer.
   Answer: We have:

$$u = \frac{dx}{dt} = 3 \qquad \text{and} \qquad a = \frac{du}{dt} = 0$$

4.1.8 Energy 118

and hence $f = ma\gamma^3 = 0$.
Comment: it is obvious that the velocity of the object is constant and hence the acceleration and force vanish. This shows that Newton's first law is also valid in Lorenz mechanics as in classical mechanics. This can also be inferred from the equation $f = ma\gamma^3$ since $f = 0$ iff $a = 0$ (since $m \ne 0$ and $\gamma \ne 0$).

3. Repeat the previous question but the equation of motion is given this time by:

$$x = \cos At$$

where A is a constant and t is time.
Answer: We have:

$$u = \frac{dx}{dt} = -A \sin At \qquad \text{and} \qquad a = \frac{du}{dt} = -A^2 \cos At$$

Hence:

$$f = ma\gamma^3 = \frac{-A^2 \cos At}{\left(1 - \frac{A^2 \sin^2 At}{c^2}\right)^{3/2}}$$

4.1.8 Energy

1. Provide more clarification about the term "total energy".
   Answer: The term "total energy", which is symbolized by E_t and is given by $E_t = \gamma mc^2$, represents the sum of rest and kinetic energy of a massive object. Hence, if this term should really represent the total energy of the object, all forms of non-kinetic energy (e.g. thermal and potential) should be included in the rest energy due to the equivalence between mass and energy.
2. Classify the rest energy, the kinetic energy and the total energy as intrinsic or extrinsic properties.
   Answer: While the rest energy is an intrinsic property since it depends only on the rest mass of the object (which should include all forms of non-kinetic energy), the kinetic and total energies are frame dependent since they depend on the speed of the object relative to the observer and hence they are extrinsic properties.
3. Discuss the impact of the Poincare mass-energy equivalence relation (see § 4.3.2 and § 5.3.2) on the use of the mass and energy units.
   Answer: Because of the mass-energy equivalence according to the Poincare relation $E_0 = mc^2$, energy units like MeV are commonly used (especially in particle physics) as mass units with $1/c^2$ being a scale factor, e.g. the mass of electron is given as 0.511 MeV/c^2. Conversely, mass units (e.g. atomic mass unit which is abbreviated as amu) may be used as energy units in some applications with c^2 being a scale factor.
4. A mother particle of mass $m_m = 400\,\text{MeV}/c^2$, which is at rest in the laboratory frame, decays into two daughter particles each of mass $m_d = 150\,\text{MeV}/c^2$. What is the kinetic energy of each one of the daughter particles?
   Answer: It is obvious that the rest energy of the mother particle is $E_m = 400\,\text{MeV}$

4.1.8 Energy

while the rest energy of each one of the daughter particles is $E_d = 150\,\text{MeV}$. So, according to the conservation of total energy we have:

$$\begin{aligned} E_m &= 2E_d + E_k \\ 400 &= 300 + E_k \\ E_k &= 100\,\text{MeV} \end{aligned}$$

where E_k is the sum of the kinetic energies of the daughter particles. Now, due to the conservation of momentum, the total momentum of the daughter particles should be zero because the mother particle is at rest, and hence if we note that the mass of the two daughter particles is identical, then the two daughter particles should have velocities that are equal in magnitude and opposite in direction. Therefore, the two daughter particles should have identical kinetic energy because they have identical mass and identical speed. This means that the kinetic energy of each daughter particle is $50\,\text{MeV}$.

5. Find a general relation between the Lorentzian momentum and the Lorentzian total energy. Also, find a general relation between the Lorentzian momentum and the Lorentzian rest and kinetic energies.
Answer: We have:

$$p = \gamma m u = \gamma m c^2 \frac{u}{c^2} = E_t \frac{u}{c^2} = E_t \frac{\beta}{c}$$

From this relation, we obtain:

$$p = E_t \frac{u}{c^2} = (E_0 + E_k) \frac{u}{c^2} = (E_0 + E_k) \frac{\beta}{c}$$

6. The momentum of a proton is $p = 100\,\text{MeV}/c$. What is its kinetic energy?
Answer: The mass of proton is $m_p \simeq 938.27\,\text{MeV}/c^2$. Hence:

$$\begin{aligned} p &= \gamma m u \\ \gamma u &= \frac{p}{m} \\ \gamma u &= \frac{100\,\text{MeV}/c}{938.27\,\text{MeV}/c^2} \\ \gamma u &= \frac{100 c}{938.27} \\ \gamma \beta &= \frac{100}{938.27} \\ \sqrt{\gamma^2 - 1} &= \frac{100}{938.27} \\ \gamma &= \sqrt{\left(\frac{100}{938.27}\right)^2 + 1} \simeq 1.005664 \end{aligned}$$

where in the sixth step we used the identity $\gamma\beta = \sqrt{\gamma^2 - 1}$ which was proved in § 1.4. Hence, the kinetic energy of the proton is:

$$E_k = m_p c^2 (\gamma - 1) \simeq (938.27\,\text{MeV}/c^2) \times c^2 \times (1.005664 - 1) \simeq 5.31\,\text{MeV}$$

4.1.8 Energy

7. What is the speed, according to classical and Lorentz mechanics, of an electron whose kinetic energy is 10 MeV? Comment on the results.
   Answer: The mass of electron is $m_e \simeq 0.511 \, \text{MeV}/c^2$ and hence its rest energy is $E_0 \simeq 0.511 \, \text{MeV}$.
   According to classical mechanics:[73]

$$\begin{aligned} E_k &= \frac{1}{2} m_e u^2 \\ u &= \sqrt{\frac{2 E_k}{m_e}} \\ u &= \sqrt{\frac{2 \times 10 \, \text{MeV}}{0.511 \, \text{MeV}/c^2}} \\ u &= c\sqrt{\frac{20}{0.511}} \simeq 6.256 c \end{aligned}$$

   According to Lorentz mechanics:

$$\begin{aligned} E_k &= m_e c^2 (\gamma - 1) \\ E_k &= E_0 (\gamma - 1) \\ \gamma &= \frac{E_k}{E_0} + 1 \\ \sqrt{1 - \beta^2} &= \left(\frac{E_k}{E_0} + 1\right)^{-1} \\ 1 - \beta^2 &= \left(\frac{E_k}{E_0} + 1\right)^{-2} \\ \beta &= \left[1 - \left(\frac{E_k}{E_0} + 1\right)^{-2}\right]^{1/2} \simeq 0.9988 \\ u &\simeq 0.9988 c \end{aligned}$$

   Comment: while the classical speed exceeds c, the Lorentzian speed stays below c although it becomes very close to c as it approaches c asymptotically from below.

8. Given that the mass of electron, proton and neutron in kilogram are: $m_e \simeq 9.109384 \times 10^{-31}$, $m_p \simeq 1.672622 \times 10^{-27}$ and $m_n \simeq 1.674929 \times 10^{-27}$, find their rest energies in joules.
   Answer:

$$E_{0e} = m_e c^2 \simeq 9.109384 \times 10^{-31} \times (3 \times 10^8)^2 \simeq 8.198446 \times 10^{-14} \, \text{J}$$

---

[73] Some of the following steps (which may also be found in similar exercises) are based on the mass-energy equivalence which may be seen as inappropriate to use in this classical context. However, as we will see the mass-energy equivalence may be classically obtained (see for example § 5.3.2). Moreover, this can be considered as a mathematical technique with no physical implication.

4.1.8 Energy

$$E_{0p} = m_p c^2 \simeq 1.672622 \times 10^{-27} \times \left(3 \times 10^8\right)^2 \simeq 1.505360 \times 10^{-10} \text{ J}$$
$$E_{0n} = m_n c^2 \simeq 1.674929 \times 10^{-27} \times \left(3 \times 10^8\right)^2 \simeq 1.507436 \times 10^{-10} \text{ J}$$

9. Find the Lorentzian speed of a massive object whose kinetic energy is half its rest energy.
   Answer: We have:

$$\begin{aligned}
E_k &= E_0/2 \\
E_0(\gamma - 1) &= E_0/2 \\
\gamma &= 3/2 \\
\sqrt{1 - \beta^2} &= 2/3 \\
\beta^2 &= 5/9 \\
\beta &= \sqrt{5}/3 \\
u &= \frac{\sqrt{5}}{3} c
\end{aligned}$$

10. Find the Lorentzian speed of a massive object whose classical speed is equal to the characteristic speed of light c. Solve the question by equating the classical kinetic energy to the Lorentzian kinetic energy. Comment on the result.
    Answer: The classical kinetic energy E_{kc} is given by:

$$E_{kc} = \frac{1}{2} m u^2 = \frac{1}{2} m c^2 = \frac{1}{2} E_0$$

   while the Lorentzian kinetic energy E_{kl} is given by:

$$E_{kl} = E_0(\gamma - 1)$$

   So, if we equate the classical kinetic energy to the Lorentzian kinetic energy we obtain:

$$\begin{aligned}
E_{kl} &= E_{kc} \\
E_0(\gamma - 1) &= \frac{1}{2} E_0 \\
\gamma &= \frac{3}{2} \\
u &= \frac{\sqrt{5}}{3} c
\end{aligned}$$

   where the last step is obtained from the answer of the previous question.
   Comment: the results of this question and the previous question show that an object whose classical speed is equal to the characteristic speed of light c has a Lorentzian kinetic energy that is half its rest energy, i.e. its total energy is $\frac{3}{2}$ times its rest energy. This should be obvious from the classical formula for kinetic energy $E_{kc} = \frac{1}{2} m u^2$ when $u = c$.

11. What is the condition that should be imposed on the speed if the maximum allowed relative error in the kinetic energy between the Lorentzian and classical formulae should be less than 1%?

    Answer: If we label the classical and Lorentzian kinetic energy as E_{kc} and E_{kl} then we should have:

    $$\frac{E_{kl} - E_{kc}}{E_{kl}} < 0.01$$
    $$0.99 E_{kl} < E_{kc}$$
    $$0.99 mc^2 (\gamma - 1) < 0.5 mu^2$$
    $$\gamma - 1 < \frac{50}{99}\beta^2$$
    $$\frac{1}{\sqrt{1-\beta^2}} < \frac{50}{99}\beta^2 + 1$$
    $$\frac{1}{1-\beta^2} < \left(\frac{50}{99}\beta^2 + 1\right)^2$$

    On solving this inequality analytically or numerically we obtain the condition $u \lesssim 0.1154c$.

12. What is the momentum of a neutron whose kinetic energy is 100 MeV?

    Answer: From the definition of kinetic energy we have:

    $$E_k = E_0 (\gamma - 1)$$
    $$\gamma = \frac{E_k}{E_0} + 1$$
    $$\gamma \simeq \frac{100}{939.57} + 1$$

    From the definition of momentum we have:

    $$p = \gamma m u$$
    $$= c \gamma m \beta$$
    $$= cm\sqrt{\gamma^2 - 1}$$
    $$\simeq c \times 939.57\,\mathrm{MeV}/c^2 \times \sqrt{\left(\frac{100}{939.57} + 1\right)^2 - 1}$$
    $$\simeq 444.88\,\mathrm{MeV}/c$$

    where the identity $\gamma\beta = \sqrt{\gamma^2 - 1}$ is used in the third step (see § 1.4).

4.1.9 Work

1. Find the amount of work required to accelerate an object of mass $m = 10^{-6}$ kg from $u_1 = 0.2c$ to $u_2 = 0.5c$.

Answer: We have:
$$W = mc^2 \left(\gamma_2 - \gamma_1\right) \simeq 10^{-6} \times \left(3 \times 10^8\right)^2 \times \left(\frac{1}{\sqrt{1 - 0.5^2}} - \frac{1}{\sqrt{1 - 0.2^2}}\right) \simeq 1.2067 \times 10^{10} \text{ J}$$

4.2 Physical Transformations

1. The given formulation of Lorentz transformations is based on a state of standard setting between the involved reference frames. What this means?
   Answer: According to the standard setting of reference frames which are in a state of relative motion, we have two inertial frames, O and O', with underlying rectangular Cartesian coordinate systems where O' is moving with a constant velocity v relative to O along the common x-x' axis. The main characteristics of the standard setting of inertial frames are:
   • The x-x' coordinate axes of the two frames remain coincident during the motion along the common x-x' dimension while the y-y' and z-z' coordinate axes remain parallel at all times.
   • The corresponding spatial axes (i.e. x-x', y-y' and z-z') of the two frames have the same sense of orientation, i.e. their positive and negative directions are identical.
   • There is no relative motion between the two systems along the orientation of the y-y' and z-z' coordinate axes and hence these coordinate axes remain parallel[74] in the two systems at all times. Accordingly, the second and third spatial coordinates of any event will be identical in both frames at all times.
   • The origins of the two coordinate systems coincide at the origin of time, that is $x = x' = 0$ at $t = t' = 0$. Accordingly, the two frames become identical at this instant of time since all the spatial and temporal coordinates of the two frames coincide.
2. The Lorentz spacetime coordinate transformations were proposed to replace which transformations? What is the original motive for this proposal?
   Answer: The Lorentz spacetime coordinate transformations were proposed to replace the Galilean transformations of classical mechanics. The original motive was the failure of the Michelson-Morley experiment to detect the hypothesized ether wind. This failure highlighted a number of problematic issues in classical physics such as the nature of the characteristic speed of light c and to which frame it belongs, and the failure of Maxwell's equations to be form invariant under the Galilean transformations.
3. Make a brief comparison between the Lorentzian and Galilean transformations of spatial and temporal coordinates highlighting their common features as well as their differences.
   Answer: The following are some comparison points between these sets of transformations:
   • Both sets transform spatial and temporal coordinates from one frame of reference to another frame of reference.
   • Both sets apply to inertial frames.

---
[74] Or rather: remain identical (due to lack of displacement along the y and z orientations) apart from the displacement along the x orientation.

- The Galilean transformations keep the invariance of space interval and time interval separately but not the spacetime interval, while the Lorentzian transformations keep the invariance of spacetime interval but not the space interval or the time interval.
- Both sets keep the invariance of the laws of classical mechanics. However, only the Lorentzian transformations keep the invariance of the laws of electromagnetism as represented by Maxwell's equations and their derived consequences like the electromagnetic wave equation.
- The Lorentzian transformations converge to the Galilean transformations at low speeds relative to c, i.e. $v \ll c$. Hence, the two sets become practically identical at this low-speed limit.
- The Lorentzian transformations are limited to speeds below the characteristic speed of light c, while the Galilean transformations are not restricted by any speed limit.

4. What is the significance of the convergence of Lorentz transformations to the Galilean transformations at low speeds?
Answer: The significance is that: classical mechanics is a valid approximation to Lorentz mechanics at low speeds. This is also an endorsement to the Galilean transformations as valid transformations within the limits of classical mechanics.

4.2.1 Lorentz Spacetime Coordinate Transformations

1. Write the Lorentz spacetime coordinate transformations between two inertial frames, O and O', which are in a state of standard setting with a relative velocity $v = -0.6c$.
Answer:
Transformations from O to O':

$$x' = \gamma(x - vt) = \frac{x + 0.6ct}{\sqrt{1 - 0.36}}$$
$$y' = y$$
$$z' = z$$
$$t' = \gamma\left(t - \frac{vx}{c^2}\right) = \frac{t + (0.6x/c)}{\sqrt{1 - 0.36}}$$

Transformations from O' to O:

$$x = \gamma(x' + vt') = \frac{x' - 0.6ct'}{\sqrt{1 - 0.36}}$$
$$y = y'$$
$$z = z'$$
$$t = \gamma\left(t' + \frac{vx'}{c^2}\right) = \frac{t' - (0.6x'/c)}{\sqrt{1 - 0.36}}$$

2. O and O' are two inertial frames in a state of standard setting with $v = -0.4c$. An event V_1 is observed from O to occur at $x_1 = 33.9$ and $t_1 = 2 \times 10^{-9}$. What are x'_1 and

4.2.1 Lorentz Spacetime Coordinate Transformations

t'_1? If another event V_2 occurred at $x'_2 = 60$ and $t'_2 = 8 \times 10^{-8}$, what are x_2 and t_2?

Answer: Using the Lorentz transformations from frame O to frame O' we obtain:

$$x'_1 = \gamma(x_1 - vt_1) = \frac{33.9 + 0.4c \times 2 \times 10^{-9}}{\sqrt{1 - 0.16}} \simeq 37.25 \text{ m}$$

$$t'_1 = \gamma\left(t_1 - \frac{vx_1}{c^2}\right) = \frac{2 \times 10^{-9} + 0.4 \times (33.9/c)}{\sqrt{1 - 0.16}} \simeq 5.15 \times 10^{-8} \text{ s}$$

For the second part of the question, we use the Lorentz transformations from frame O' to frame O and hence we obtain:

$$x_2 = \gamma(x'_2 + vt'_2) = \frac{60 - 0.4c \times 8 \times 10^{-8}}{\sqrt{1 - 0.16}} \simeq 54.99 \text{ m}$$

$$t_2 = \gamma\left(t'_2 + \frac{vx'_2}{c^2}\right) = \frac{8 \times 10^{-8} - 0.4 \times (60/c)}{\sqrt{1 - 0.16}} = 0$$

3. An event is observed from two inertial frames O and O', which are in a state of standard setting, to be at $\mathbf{x} = (12, 6, -21, 3)$ and $\mathbf{x}' = (10.4745, 1.3093, -21, 3)$ respectively where the temporal coordinates ct and ct' are the first. What is the relative speed between the two frames?

Answer: Taking the ratio of the Lorentz transformation of the temporal coordinate to the Lorentz transformation of the x spatial coordinate, we obtain:

$$\frac{ct'}{x'} = \frac{\gamma(ct - \beta x)}{\gamma(x - \beta ct)}$$

$$\frac{10.4745}{1.3093} = \frac{12 - 6\beta}{6 - 12\beta}$$

$$\frac{12 - 6\beta}{6 - 12\beta} \simeq 8$$

$$12 - 6\beta \simeq 48 - 96\beta$$

$$\beta \simeq \frac{36}{90}$$

$$v \simeq 0.4c$$

4. V_1 is an event seen in an inertial frame O to take place at $x_1 = \xi$, $y_1 = z_1 = 0$ and $ct_1 = \xi$, and V_2 is another event seen in O to take place at $x_2 = 3\xi$, $y_2 = z_2 = 0$ and $ct_2 = 2\xi$ where ξ is a positive number. However, V_1 and V_2 are seen in another inertial frame O' to take place at the same time. Assuming that O and O' are in a state of standard setting, what is the speed of O' relative to O? Comment on the significance of this exercise.

Answer: The temporal coordinates of the two events in frame O' are given by:

$$ct'_1 = \gamma(ct_1 - \beta x_1)$$
$$ct'_2 = \gamma(ct_2 - \beta x_2)$$

4.2.1 Lorentz Spacetime Coordinate Transformations

If these two temporal coordinates are to be equal then we should have:

$$ct'_1 = ct'_2$$
$$\gamma(ct_1 - \beta x_1) = \gamma(ct_2 - \beta x_2)$$
$$\beta x_2 - \beta x_1 = ct_2 - ct_1$$
$$\beta = \frac{ct_2 - ct_1}{x_2 - x_1}$$
$$\beta = \frac{2\xi - \xi}{3\xi - \xi}$$
$$\beta = \frac{1}{2}$$
$$v = 0.5c$$

Comment: this is an example of the relativity of simultaneity (see § 7.4) where V_1 and V_2 are simultaneous in frame O' but not in frame O.

5. O and O' are two inertial observers in a state of standard setting with $v = -0.45c$. Two events, V_1 and V_2, are seen by O to be co-positional, while they are seen by O' to be separated by a time interval of 0.002. What is the spatial separation between V_1 and V_2 according to O'? Comment on the significance of this exercise.

Answer: The x coordinates of the two events in frame O are given by:

$$x_1 = \gamma(x'_1 + vt'_1)$$
$$x_2 = \gamma(x'_2 + vt'_2)$$

If these two x coordinates are to be equal, because V_1 and V_2 are co-positional in frame O, then we should have:

$$x_1 = x_2$$
$$\gamma(x'_1 + vt'_1) = \gamma(x'_2 + vt'_2)$$
$$x'_1 - x'_2 = v(t'_2 - t'_1)$$
$$x'_1 - x'_2 = -0.45c \times 0.002$$
$$x'_2 - x'_1 = 2.7 \times 10^5 \text{ m}$$

Comment: this is an example of the relativity of co-positionality (see § 7.5) where V_1 and V_2 are co-positional in frame O but not in frame O'.

6. O and O' are two inertial frames in a state of standard setting with $v = 0.3c$. An object moves in the xz plane of frame O with a constant velocity $u = 0.4c$ where its straight path makes an angle $\theta = \pi/6$ with the positive x axis. What are the equations of motion of the object in frame O' (i.e. x' and z' as functions of t')?

Answer: The equations of motion in frame O are given by:

$$x = u\left(\cos\frac{\pi}{6}\right)t = \frac{\sqrt{3}}{2}ut = \sqrt{0.12}ct$$
$$z = u\left(\sin\frac{\pi}{6}\right)t = \frac{1}{2}ut = 0.2ct$$

4.2.1 Lorentz Spacetime Coordinate Transformations

On applying the Lorentz transformations to the x equation, we obtain for x':
$$\begin{aligned}
x &= \sqrt{0.12}\,ct \\
\gamma\left(x' + vt'\right) &= \sqrt{0.12}\,\gamma\left(ct' + \beta x'\right) \\
x' + vt' &= \sqrt{0.12}\,ct' + \sqrt{0.12}\,\beta x' \\
x' - 0.3\sqrt{0.12}\,x' &= \sqrt{0.12}\,ct' - 0.3ct' \\
x' &= \frac{\sqrt{0.12} - 0.3}{1 - 0.3\sqrt{0.12}}\,ct' \\
x' &= Act'
\end{aligned}$$

where $A = \frac{\sqrt{0.12}-0.3}{1-0.3\sqrt{0.12}}$. Similarly, we obtain for z':

$$\begin{aligned}
z &= 0.2\,ct \\
z' &= 0.2\gamma\left(ct' + \beta x'\right) \\
z' &= 0.2\gamma\left(ct' + \beta Act'\right) \\
z' &= 0.2\gamma\left(ct' + 0.3 Act'\right) \\
z' &= 0.2\left(1 + 0.3A\right)\gamma ct' \\
z' &= \frac{0.2 + 0.06A}{\sqrt{0.91}}\,ct' \\
z' &= Bct'
\end{aligned}$$

where $B = \frac{0.2+0.06A}{\sqrt{0.91}}$. We note that $y' = y = 0$.

7. An inertial observer O measures the spatial separation between two events, V_1 and V_2, to be $x_2 - x_1 = 10^8$ and the time interval to be $t_2 - t_1 = 1$. What is the proper time interval between V_1 and V_2?

   Answer: Because the spatial separation between V_1 and V_2 in O frame is not zero, then his time interval is not proper. Hence, the proper time interval belongs to another inertial observer O' who moves with a velocity v relative to O where in O' frame the two events are observed co-positional (i.e. $x'_2 - x'_1 = 0$), that is:

$$\begin{aligned}
x'_2 - x'_1 &= \gamma\left(x_2 - vt_2\right) - \gamma\left(x_1 - vt_1\right) \\
0 &= \gamma\left[\left(x_2 - x_1\right) - v\left(t_2 - t_1\right)\right] \\
0 &= \left(x_2 - x_1\right) - v\left(t_2 - t_1\right) \\
v &= \frac{x_2 - x_1}{t_2 - t_1} = 10^8
\end{aligned}$$

Accordingly, the proper time interval (i.e. $t'_2 - t'_1$) is given by:

$$\begin{aligned}
t'_2 - t'_1 &= \gamma\left(t_2 - \frac{vx_2}{c^2}\right) - \gamma\left(t_1 - \frac{vx_1}{c^2}\right) \\
&= \gamma\left[\left(t_2 - t_1\right) - \frac{v}{c^2}\left(x_2 - x_1\right)\right] \\
&= \frac{1 - \left(10^8/c^2\right)\times 10^8}{\sqrt{1 - \left(10^8/c\right)^2}} \\
&\simeq 0.9428\,\text{s}
\end{aligned}$$

The proper time interval can also be obtained more easily from the time dilation formula, that is:

$$t'_2 - t'_1 = \frac{t_2 - t_1}{\gamma} = 1 \times \sqrt{1 - (10^8/c)^2} \simeq 0.9428\,\text{s}$$

4.2.2 Velocity Transformations

1. An Object is seen by an inertial observer O to have a velocity $\mathbf{u} = (u_x, u_y, u_z) = (0.3c, 0.1c, 0.4c)$. What is the speed of the object as seen by another inertial observer O' who is in a state of standard setting with O where $v = 0.2c$?
   Answer: Using the Lorentzian velocity transformation formulae from O to O', we have:

$$u'_x = \frac{u_x - v}{1 - \frac{vu_x}{c^2}} = \frac{0.3c - 0.2c}{1 - 0.2 \times 0.3} \simeq 0.1064c$$

$$u'_y = \frac{u_y}{\gamma\left(1 - \frac{vu_x}{c^2}\right)} = \frac{0.1c\sqrt{1 - 0.2^2}}{1 - 0.2 \times 0.3} \simeq 0.1042c$$

$$u'_z = \frac{u_z}{\gamma\left(1 - \frac{vu_x}{c^2}\right)} = \frac{0.4c\sqrt{1 - 0.2^2}}{1 - 0.2 \times 0.3} \simeq 0.4169c$$

   Hence, the speed of the object as seen by O' is:

$$u' = \sqrt{(u'_x)^2 + (u'_y)^2 + (u'_z)^2} \simeq 0.4427c$$

2. An elementary (mother) particle decays into two (daughter) particles of equal speed. In the rest frame of the mother particle, the speed of each one of the daughter particles is $0.75c$. What are the velocities of the daughter particles in a frame in which the mother particle is moving at a speed of $0.5c$? Assume in your answer that the daughter particles move along the same orientation as the motion of the mother particle.
   Answer: We label with O the frame from which the mother particle is seen to have a speed of $0.5c$, while we label with O' the frame in which the mother particle is at rest. Moreover, we assume that all motions are oriented along the x axis. So, we have two inertial frames, O and O', which are in a state of standard setting with $v = 0.5c$ where in frame O' the two daughter particles have $u'_x = \pm 0.75c$ and $u'_y = u'_z = 0$. Accordingly, we have:

$$u_x = \frac{u'_x + v}{1 + \frac{vu'_x}{c^2}} = \frac{\pm 0.75c + 0.5c}{1 + \frac{0.5c \times (\pm 0.75)c}{c^2}} = \frac{\pm 0.75 + 0.5}{1 \pm 0.375}c$$

   while $u_y = u_z = 0$. So, the velocities of the daughter particles are:

$$u_{x1} \simeq 0.9091c \qquad \text{and} \qquad u_{x2} = -0.4c$$

   i.e. one of the daughter particles will be seen from O to be moving with a speed of about $0.9091c$ in the positive x direction while the other daughter particle will be seen

4.2.2 Velocity Transformations

to be moving with a speed of $0.4c$ in the negative x direction. If the mother particle is moving in the negative x direction (i.e. $v = -0.5c$) then the signs of the velocities of the daughter particles should be reversed.

3. Repeat the previous exercise but this time assume that the daughter particles, as seen from the rest frame of the mother particle, move in a perpendicular direction to the direction of the relative motion between the two frames.
   Answer: Everything is the same as in the answer of the previous exercise, except that we now assume that the motion of the daughter particles are in the orientation of the y' axis and hence we have $u'_x = u'_z = 0$ and $u'_y = \pm 0.75c$. Accordingly, we have:

$$u_x = \frac{u'_x + v}{1 + \frac{vu'_x}{c^2}} = \frac{0 + 0.5c}{1 + 0} = 0.5c$$

$$u_y = \frac{u'_y}{\gamma\left(1 + \frac{vu'_x}{c^2}\right)} = \frac{\pm 0.75c\sqrt{1 - 0.5^2}}{1 + 0} = \pm 0.75c\sqrt{1 - 0.5^2} \simeq \pm 0.6495c$$

$$u_z = 0$$

So, the daughter particles move in the xy plane with a speed u given by:

$$u = \sqrt{(u_x)^2 + (u_y)^2 + (u_z)^2} \simeq c\sqrt{0.5^2 + (\pm 0.6495)^2 + 0^2} \simeq 0.8197c$$

and their straight paths make angles $\pm \theta$ with the positive x axis of the O frame where θ is given by:

$$\theta = \arctan\frac{|u_y|}{u_x} \simeq 0.9147\,\text{rad} \simeq 52.41°$$

4. O and O' are two inertial observers in a state of standard setting with $v = 0.7c$. An object moves in the xy plane of O frame with a constant speed $u = 0.15c$ where its straight path makes an angle $\theta = \pi/3$ with the positive x axis, i.e. the x and y components of its constant velocity are positive. What is the velocity of the object in O' frame? Find this velocity once as components along the coordinate axes and once as speed and direction.
   Answer: In O frame we have:

$$u_x = u\cos\frac{\pi}{3} = 0.15c \times \frac{1}{2} = 0.075c$$

$$u_y = u\sin\frac{\pi}{3} = 0.15c \times \frac{\sqrt{3}}{2} = \sqrt{0.016875}c$$

$$u_z = 0$$

Using the Lorentzian velocity transformations, the velocity components in O' frame are:

$$u'_x = \frac{u_x - v}{1 - \frac{vu_x}{c^2}} = \frac{0.075c - 0.7c}{1 - (0.7 \times 0.075)} \simeq -0.6596c$$

$$u'_y = \frac{u_y}{\gamma\left(1 - \frac{vu_x}{c^2}\right)} = \frac{\sqrt{0.016875}c\sqrt{1 - 0.7^2}}{1 - (0.7 \times 0.075)} \simeq 0.0979c$$

$$u'_z = 0$$

4.2.2 Velocity Transformations

Hence, the velocity of the object in O' frame in terms of components is:
$$\mathbf{u}' = \left(u'_x, u'_y, u'_z\right) \simeq (-0.6596, 0.0979, 0)\, c$$

Accordingly, the speed of the object in O' frame is:
$$u' = \sqrt{\left(u'_x\right)^2 + \left(u'_y\right)^2 + \left(u'_z\right)^2} \simeq c\sqrt{(-0.6596)^2 + 0.0979^2 + 0^2} \simeq 0.6669c$$

and it moves in the $x'y'$ plane where the angle θ' that its straight path makes with the positive x' axis in O' frame is:
$$\theta' = \arctan\frac{u'_y}{u'_x} \simeq 2.9942\,\text{rad} \simeq 171.56°$$

5. The observed speed of light in transparent material media, like air and water, is given by c/n where c is the characteristic speed of light in free space and n is the refractive index of the particular medium which is assumed to be in a standstill state. According to the Fizeau formula, which he derived[75] from his experiments in which he measured the speed of light in running water, the observed speed of light c_m in a medium that is moving uniformly with speed u_m relative to the observer is given by:
$$c_m = \frac{c}{n} + k u_m$$

where k is the dragging coefficient which is a medium-dependent parameter. In the mid 19th century, Fizeau found that the dragging coefficient of water is $k_w \simeq 0.44$. Try to justify this finding (both formula and $k_w \simeq 0.44$) by using the Lorentzian velocity transformations. Also, comment on the implication of this finding.

Answer: If we label the laboratory frame with O and the water frame with O' and consider all the motions to be along the x axis, then we have two inertial frames, O and O', which are in a state of standard setting with $v = u_m$ and accordingly $u'_x = c/n$ and $u_x = c_m$ where c_m is the velocity of light as seen from frame O. Using the Lorentzian velocity transformation from frame O' to frame O, we obtain:

$$c_m \equiv u_x = \frac{u'_x + v}{1 + \frac{v u'_x}{c^2}} = \frac{(c/n) + u_m}{1 + \frac{u_m(c/n)}{c^2}} = \frac{(c/n) + u_m}{1 + \frac{u_m}{cn}} = \left(\frac{c}{n} + u_m\right)\left(1 + \frac{u_m}{cn}\right)^{-1}$$

Now, since $n > 1$ and hence $u_m \ll cn$, we use the power series expansion approximation to obtain:
$$\left(1 + \frac{u_m}{cn}\right)^{-1} \simeq 1 - \frac{u_m}{cn}$$

where the quadratic and higher terms in $\frac{u_m}{cn}$ are discarded because they are negligible. Hence, we obtain:

$$c_m \simeq \left(\frac{c}{n} + u_m\right)\left(1 - \frac{u_m}{cn}\right) = \frac{c}{n} - \frac{u_m}{n^2} + u_m - \frac{u_m^2}{cn} = \frac{c}{n} + u_m\left(1 - \frac{1}{n^2}\right) - \frac{u_m^2}{cn}$$

---
[75] We should also refer to Fresnel in this regard (see § 2.6).

4.2.2 Velocity Transformations

Again, since $u_m^2 \ll cn$, we discard the last term and hence we obtain:

$$c_m \simeq \frac{c}{n} + u_m\left(1 - \frac{1}{n^2}\right) = \frac{c}{n} + ku_m$$

which is the Fizeau formula with $k = (1 - 1/n^2)$. For water we have $n \simeq 4/3$ and hence:

$$k_w = 1 - \frac{1}{n^2} \simeq 1 - \frac{9}{16} = \frac{7}{16} \simeq 0.44$$

which justifies the value found by Fizeau for k_w.

Comment: the results of Fizeau are considered as support to Lorentz mechanics in general and to the Lorentzian velocity transformations in particular. These findings may also be used to support the proposition that light behaves like material waves in classical mechanics where its characteristic speed is determined with respect to the medium of propagation (whether ether or water).[76] In the Fizeau experiment, the light source was at rest in the laboratory frame. However, the physical laws of propagation of light in free space and in transparent material media may be fundamentally different. Since our primary interest in Lorentz mechanics is the speed of light in free space, we do not discuss this issue further.

6. O and O' are two inertial observers in a state of standard setting with $v = 0.5c$. O' sent a light ray in the positive y' direction. What is the speed and direction of this signal as seen from O frame?

Answer: According to the given information, we have $u'_x = u'_z = 0$ and $u'_y = c$. Using the Lorentz velocity transformations, the velocity components of the light signal in O frame are:

$$u_x = \frac{u'_x + v}{1 + \frac{vu'_x}{c^2}} = \frac{0 + 0.5c}{1 + 0} = 0.5c$$

$$u_y = \frac{u'_y}{\gamma\left(1 + \frac{vu'_x}{c^2}\right)} = \frac{c\sqrt{1 - 0.5^2}}{1 + 0} = c\sqrt{0.75}$$

$$u_z = 0$$

Hence, the speed in O frame is:

$$u = \sqrt{u_x^2 + u_y^2 + u_z^2} = c\sqrt{0.25 + 0.75 + 0} = c$$

---

[76] We note that according to the drift theory of Fresnel, which the Fizeau experiment is based upon, it is conjectured that in the propagation of light (which is supposed to be facilitated by ether even in material media) through a moving material medium the medium will drag the ether with it. Therefore, the Fizeau experiment was originally seen as verification to the drift theory and hence the existence of ether. However, there are many debates about these issues with no argument that can be seen as completely conclusive as they generally rely upon incompatible theoretical frameworks and interpretations. Therefore, it is not very useful to go through these details. The interested reader is referred to the wider literature of this subject although we strongly advise against spending valuable time and effort on pursuing such irresolvable controversies instead of focusing on the big issues.

4.2.2 Velocity Transformations

and the angle that this light ray makes with the positive x axis of O frame is:

$$\theta = \arctan \frac{u_y}{u_x} = \frac{\pi}{3} \text{ rad} = 60°$$

7. O and O' are two inertial observers in a state of standard setting with $v = 0.3c$. A light signal is emitted by O' in a direction that makes an angle $\theta' = 45°$ with the positive x' axis where this signal is contained in the $x'y'$ plane. Find the velocity components of this signal and its speed and direction as seen from O frame. Comment on the result.
   Answer: In O' frame we have:

   $$\begin{aligned} u'_x &= c'_x = c\cos\frac{\pi}{4} = \frac{c}{\sqrt{2}} \\ u'_y &= c'_y = c\sin\frac{\pi}{4} = \frac{c}{\sqrt{2}} \\ u'_z &= 0 \end{aligned}$$

   From the Lorenz velocity transformations, the velocity components in O frame are:

   $$u_x = c_x = \frac{u'_x + v}{1 + \frac{vu'_x}{c^2}} = \frac{(c/\sqrt{2}) + 0.3c}{1 + \frac{0.3c \times (c/\sqrt{2})}{c^2}} = \frac{(1/\sqrt{2}) + 0.3}{1 + 0.3(1/\sqrt{2})} c \simeq 0.8309c$$

   $$u_y = c_y = \frac{u'_y}{\gamma\left(1 + \frac{vu'_x}{c^2}\right)} = \frac{(c/\sqrt{2})\sqrt{1 - 0.3^2}}{1 + \frac{0.3c \times (c/\sqrt{2})}{c^2}} = \frac{(1/\sqrt{2})\sqrt{1 - 0.3^2}}{1 + 0.3(1/\sqrt{2})} c \simeq 0.5565c$$

   $$u_z = 0$$

   Hence, the speed of the signal in O frame is:

   $$u = \sqrt{u_x^2 + u_y^2 + u_z^2} = \sqrt{c_x^2 + c_y^2 + 0^2} = c$$

   and it makes an angle θ with the positive x axis which is given by:

   $$\theta = \arctan \frac{u_y}{u_x} = \arctan \frac{c_y}{c_x} \simeq 0.5902 \text{ rad} \simeq 33.81°$$

   Comment: in both frames, the signal has the same observed speed which is the characteristic speed of light c. However, the direction of the signal, as seen by O and O' relative to their x-x' axis, is different. Accordingly, although the observed speed of light is the same in both frames, the velocity of light is different.

8. Referring to the previous question, justify the relation $\theta < \theta'$ with some graphic illustration.
   Answer: The x component of the velocity of light in O frame has increased, due to the relative motion of O' frame with speed v, and hence the y component of the velocity of light in O frame should be reduced from its value in O' frame to keep the speed of light (which is the magnitude of its velocity) constant. This explains why θ is smaller than θ'. The situation is illustrated in Figure 9 where the velocity of light in O frame is

4.2.2 Velocity Transformations   133

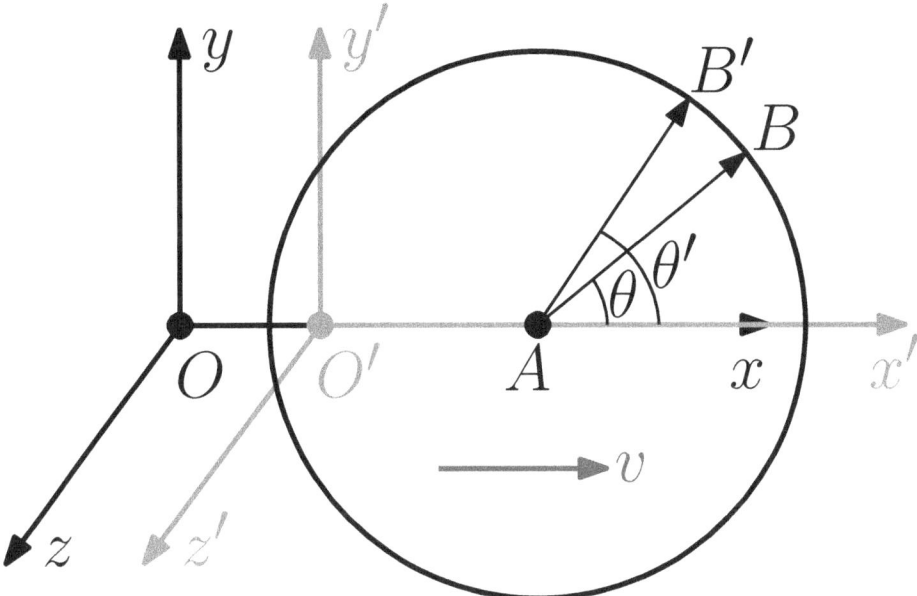

Figure 9: The velocity of light signal as seen from two inertial frames O and O' which are in a state of standard setting with relative velocity v.

represented by the vector AB while the velocity of light in O' frame is represented by the vector AB'. Both these vectors are radii of the shown circle and hence they have the same magnitude, which is the characteristic speed of light c, although they have different directions. This means that although the spacetime of Lorentz mechanics is distorted due to the motion, the speed of light keeps its value as a universal constant. In fact, the invariance of the speed of light can be explained by this distortion (i.e. the contraction of spacetime under the influence of motion), although other factors (i.e. the added velocity component and using c for calibration) also contribute to this invariance. More clarifications about these issues will come later.

9. Discuss the implications of the previous exercises about the speed and velocity of light in inertial frames.
Answer: The obvious implication of the previous exercises about the propagation of light signals in inertial frames is that while the velocity of light is frame dependent since its direction depends on the observer, the speed of light (i.e. the magnitude of the velocity of light) is frame independent since it keeps its characteristic value c in all inertial frames and hence it is value-invariant under the Lorentz transformations across all inertial frames. However, this invariance should be linked to the spacetime contraction by the motion and other factors that determine the interpretation and justification of this invariance. The previous exercises also indicate that the propagation of light in free space does not follow a classical wave propagation model (see § 1.13.2 and § 1.14) but it follows a projectile propagation model (see § 1.13.1 and § 1.14) where the velocity of light has a component from the velocity of its source and hence it depends on the motion of the source. However, we should describe this model as Lorentzian

4.2.3 Velocity Composition

projectile propagation model rather than classical projectile model because unlike the classical projectile model where the speed is frame dependent the Lorentzian model keeps the invariance of speed across frames. These issues will be detailed later.

4.2.3 Velocity Composition

1. Make a comparison between the velocity transformation formulae and the velocity composition formulae.
Answer: Some valid points in this comparison could be:
• The two sets are essentially the same and hence they represent rules for transforming velocity between inertial frames. However, as indicated earlier the velocity transformation formulae are conceptually more appropriate for transforming one velocity from one inertial frame to another inertial frame while the composition formulae are conceptually more appropriate for composing the velocity of two objects from a given inertial frame where one of the objects is seen as an inertial frame.
• Both sets are based on a state of standard setting between the frames of observation.
• Each set has its advantages and disadvantages and hence one or the other may be more suitable to use in particular cases or contexts. For example, the velocity composition formulae may have more complex symbolic structure but they are more clear and intuitive and hence they are generally easier to automate and use. Also there is no primed and unprimed symbols in the composition formulae and hence the situation between frames is more symmetric and the symbols are generally less confusing. On the other hand, the velocity transformation formulae are more clear about the relative velocity between the two frames of observation, which is symbolized distinctively with v, whereas in the composition formulae special care is needed to distinguish this velocity when the situation involves motions in more than one dimension (i.e. when the y or z formulae are needed).

2. We have two persons: person A who is standing on the street and watching a bus moving along a straight line with a constant velocity $v_B = 10$, and person C who is on the bus where C, according to A, is moving in the same direction as the bus. According to the bus rest frame, C is moving with a constant velocity $v_C = 2$. What is the velocity of C in the frame of A according to (a) classical mechanics and (b) Lorentz mechanics?
Answer: All the motions are in a single direction and hence the x velocity composition formula should be used. If we refer to A, bus and C by the subscript indices 1, 2 and 3 then we have: $u_{x21} = 10$, $u_{x32} = 2$ and u_{x31} is the required velocity of C relative to A. Accordingly:
(a) Using the Galilean velocity composition formula, we have:

$$u_{x31} = u_{x32} + u_{x21} = 2 + 10 = 12$$

(b) Using the Lorentzian velocity composition formula, we have:

$$u_{x31} = \frac{u_{x32} + u_{x21}}{1 + \frac{u_{x32}u_{x21}}{c^2}} = \frac{2 + 10}{1 + \frac{2 \times 10}{c^2}} \simeq 12.0$$

4.2.3 Velocity Composition

3. Repeat the previous question but now $v_B = 0.5c$ and $v_C = 0.2c$. Compare your results with the results of the previous question and comment.
   Answer: We use the same labeling as in the answer of the previous question and hence $u_{x21} = 0.5c$ and $u_{x32} = 0.2c$. Accordingly:
   (a) Using the Galilean velocity composition formula, we have:
   $$u_{x31} = u_{x32} + u_{x21} = 0.2c + 0.5c = 0.7c$$
   (b) Using the Lorentzian velocity composition formula, we have:
   $$u_{x31} = \frac{u_{x32} + u_{x21}}{1 + \frac{u_{x32}u_{x21}}{c^2}} = \frac{0.2c + 0.5c}{1 + 0.1} \simeq 0.6364c$$
   Comment: we see that at low speeds (i.e. $u_{x32}u_{x21} \ll c^2$), as in the previous question, the Galilean formula is very good approximation to the Lorentzian formula and hence the Galilean and Lorentzian results are virtually identical, while at high speeds (i.e. where $u_{x32}u_{x21}$ is a significant fraction of c^2), as in the present question, the Galilean result departs significantly from the Lorentzian result where the percentage relative difference between the Galilean and Lorentzian results is about 9.1%.

4. Repeat the last question but now v_B and v_C are in opposite directions.
   Answer: We use the same labeling as in the answer of the previous questions. The only difference will be that u_{x21} is opposite in sign to u_{x32} and hence either $u_{x21} = -0.5c$ and $u_{x32} = 0.2c$ or $u_{x21} = 0.5c$ and $u_{x32} = -0.2c$, that is:
   (a) The Galilean velocity:
   $$u_{x31} = u_{x32} + u_{x21} = +0.2c - 0.5c = -0.3c$$
   $$u_{x31} = u_{x32} + u_{x21} = -0.2c + 0.5c = +0.3c$$
   (b) The Lorentzian velocity:
   $$u_{x31} = \frac{u_{x32} + u_{x21}}{1 + \frac{u_{x32}u_{x21}}{c^2}} = \frac{+0.2c - 0.5c}{1 - 0.1} \simeq -0.3333c$$
   $$u_{x31} = \frac{u_{x32} + u_{x21}}{1 + \frac{u_{x32}u_{x21}}{c^2}} = \frac{-0.2c + 0.5c}{1 - 0.1} \simeq +0.3333c$$
   The magnitude of the percentage relative difference between the Galilean and Lorentzian results is about 11.1% where, unlike the previous exercise, the Galilean is now smaller in magnitude than the Lorentzian.

5. Show that when u_{x21} or/and u_{x32} in the Lorentz velocity composition formula for the x dimension approach c, u_{x31} also approaches c. What are the regions on the Lorentzian counter plot of u_{x31} that represent these cases?
   Answer: We have three cases: only $u_{x21} \to c$, only $u_{x32} \to c$, and both $u_{x21} \to c$ and $u_{x32} \to c$, that is:
   $$u_{x31} = \frac{u_{x32} + u_{x21}}{1 + \frac{u_{x32}u_{x21}}{c^2}} \simeq \frac{u_{x32} + c}{1 + \frac{u_{x32}c}{c^2}} = \frac{u_{x32} + c}{c + u_{x32}}c = c$$

4.2.3 Velocity Composition

$$u_{x31} = \frac{u_{x32} + u_{x21}}{1 + \frac{u_{x32}u_{x21}}{c^2}} \simeq \frac{c + u_{x21}}{1 + \frac{cu_{x21}}{c^2}} = \frac{c + u_{x21}}{c + u_{x21}} c = c$$

$$u_{x31} = \frac{u_{x32} + u_{x21}}{1 + \frac{u_{x32}u_{x21}}{c^2}} \simeq \frac{c + c}{1 + \frac{c^2}{c^2}} = \frac{2}{2}c = c$$

So, in all three cases u_{x31} also approaches c.

The regions on the Lorentzian counter plot of u_{x31} that represent these cases are respectively: the right edge, the upper edge, and the upper right corner.

6. Show that when u_{x21} or/and u_{x32} in the Lorentz velocity composition formula for the y dimension approach c, then u_{y31} should approach zero. Comment on the results.
Answer: We have three cases:
(a) Only u_{x21} approaches c, and hence we have:

$$u_{y31} = \frac{u_{y32}}{\gamma\left(1 + \frac{u_{x21}u_{x32}}{c^2}\right)} = \frac{u_{y32}\sqrt{1 - (u_{x21}/c)^2}}{1 + \frac{u_{x21}u_{x32}}{c^2}} \simeq \frac{u_{y32}\sqrt{1 - (c/c)^2}}{1 + \frac{cu_{x32}}{c^2}} = \frac{u_{y32}\sqrt{1 - 1}}{1 + \frac{u_{x32}}{c}} = 0$$

Comment: this result is a consequence of the issue that no physical speed should exceed c according to the formalism of Lorentz mechanics.[77] The reason is that when $u_{x21} \to c$ then $u_{x31} \to c$ as we found in the previous exercise and hence the other components of the velocity $\mathbf{u}_{31}$ should approach zero to avoid $|\mathbf{u}_{31}|$ exceeding the speed of light. More formally, we have:

$$|\mathbf{u}_{31}| = \sqrt{(u_{x31})^2 + (u_{y31})^2 + (u_{z31})^2}$$
$$\simeq \sqrt{c^2 + (u_{y31})^2 + (u_{z31})^2}$$
$$= c\sqrt{1 + (u_{y31}/c)^2 + (u_{z31}/c)^2}$$

So, if $|\mathbf{u}_{31}| \to c$ then we should have $u_{y31} \to 0$ and $u_{z31} \to 0$. We note that this case is represented by the right edge of the Lorentzian counter plot of u_{y31} assuming $u_{x32} = 0$.
(b) Only u_{x32} approaches c. In this case $u_{y32} \to 0$ and hence we have:

$$u_{y31} = \frac{u_{y32}}{\gamma\left(1 + \frac{u_{x21}u_{x32}}{c^2}\right)} = \frac{u_{y32}\sqrt{1 - (u_{x21}/c)^2}}{1 + \frac{u_{x21}u_{x32}}{c^2}} \simeq \frac{0 \times \sqrt{1 - (u_{x21}/c)^2}}{1 + \frac{u_{x21}c}{c^2}} = 0$$

Comment: again, this result is a consequence of the issue that no speed should exceed c according to Lorentz mechanics. The reason is that when $u_{x32} \to c$ then the other components of the velocity $\mathbf{u}_{32}$ should approach zero to avoid $|\mathbf{u}_{32}|$ exceeding c. More formally, we have:

$$|\mathbf{u}_{32}| = \sqrt{(u_{x32})^2 + (u_{y32})^2 + (u_{z32})^2}$$

---
[77] We mean within the given conditions that we discussed earlier and will be discussed further in the future.

4.2.3 Velocity Composition

$$\simeq \sqrt{c^2 + (u_{y32})^2 + (u_{z32})^2}$$
$$= c\sqrt{1 + (u_{y32}/c)^2 + (u_{z32}/c)^2}$$

So, if $|\mathbf{u}_{32}| \to c$ then we should have $u_{y32} \to 0$ and $u_{z32} \to 0$ and hence u_{y31} will vanish according to its formula.

(c) Both u_{x21} and u_{x32} approach c and hence we have:

$$u_{y31} = \frac{u_{y32}}{\gamma\left(1 + \frac{u_{x21}u_{x32}}{c^2}\right)} = \frac{u_{y32}\sqrt{1 - (u_{x21}/c)^2}}{1 + \frac{u_{x21}u_{x32}}{c^2}} \simeq \frac{0 \times \sqrt{1 - (c/c)^2}}{1 + \frac{c \times c}{c^2}} = 0$$

Comment: this result should be obvious because we have two reasons for u_{y31} to approach zero.

We note that all these results equally apply to the Lorentz velocity composition formula for the z dimension because the formulae for u_{y31} and u_{z31} are identical apart from the y and z subscripts.

7. Show that when u_{y32} in the Lorentz velocity composition formula for the y dimension approaches c, then u_{y31} can take any value between zero and c (i.e. $0 \le u_{y31} \le c$).[78]
   Answer: When the velocity $u_{y32} \to c$ then $u_{x32} \to 0$ because $|\mathbf{u}_{32}|$ cannot exceed c as demonstrated in the previous exercises. Hence, the u_{y31} formula becomes:

$$u_{y31} = \frac{u_{y32}}{\gamma\left(1 + \frac{u_{x21}u_{x32}}{c^2}\right)} \simeq \frac{c\sqrt{1 - (u_{x21}/c)^2}}{1 + 0} = c\sqrt{1 - (u_{x21}/c)^2}$$

Now, since $0 \le u_{x21} \le c$ because it is independent from u_{x32} and u_{y32} then $\sqrt{1 - (u_{x21}/c)^2}$ will be between 1 (corresponding to $u_{x21} = 0$) and 0 (corresponding to $u_{x21} = c$) and hence u_{y31} will be between c and 0.

8. Build a table showing u_{x31} as a function of the velocities u_{x21} and u_{x32} at and between the two limits of these velocities: zero and c.
   Answer:

	$u_{x32} = 0$	$0 < u_{x32} < c$	$u_{x32} = c$
$u_{x21} = 0$	0	u_{x32}	c
$0 < u_{x21} < c$	u_{x21}	$\frac{u_{x32}+u_{x21}}{1+\frac{u_{x32}u_{x21}}{c^2}}$	c
$u_{x21} = c$	c	c	c

9. Build a table showing u_{y31} as a function of the velocities u_{x21} and u_{x32} at and between the two limits of these velocities: zero and c.
   Answer:

---

[78] We note that in contexts like this, symbols like u_{y32} and u_{y31} may represent the magnitude of velocity. This is inline with our previously stated convention (see § 1.5) that symbols like these may represent 1D velocity and may represent speed.

4.2.3 Velocity Composition

	$u_{x32}=0$	$0<u_{x32}<c$	$u_{x32}=c$
$u_{x21}=0$	u_{y32}	u_{y32}	0
$0<u_{x21}<c$	$u_{y32}\sqrt{1-(u_{x21}/c)^2}$	$\dfrac{u_{y32}\sqrt{1-(u_{x21}/c)^2}}{1+\frac{u_{x21}u_{x32}}{c^2}}$	0
$u_{x21}=c$	0	0	0

10. Build a table showing u_{y31} as a function of the velocities u_{x21} and u_{y32} at and between the two limits of these velocities: zero and c.
    Answer:

	$u_{y32}=0$	$0<u_{y32}<c$	$u_{y32}=c$
$u_{x21}=0$	0	u_{y32}	c
$0<u_{x21}<c$	0	$\dfrac{u_{y32}\sqrt{1-(u_{x21}/c)^2}}{1+\frac{u_{x21}u_{x32}}{c^2}}$	$c\sqrt{1-(u_{x21}/c)^2}$
$u_{x21}=c$	0	0	0

11. Build a table showing u_{y31} as a function of the velocities u_{x32} and u_{y32} at and between the two limits of these velocities: zero and c.
    Answer: (NA $\equiv$ Not Applicable)

	$u_{x32}=0$	$0<u_{x32}<c$	$u_{x32}=c$
$u_{y32}=0$	0	0	0
$0<u_{y32}<c$	$u_{y32}\sqrt{1-(u_{x21}/c)^2}$	$\dfrac{u_{y32}\sqrt{1-(u_{x21}/c)^2}}{1+\frac{u_{x21}u_{x32}}{c^2}}$	NA
$u_{y32}=c$	$c\sqrt{1-(u_{x21}/c)^2}$	NA	NA

12. Having three objects or observers (O, O_1 and O_2), the formula:

$$u_{x31} = \frac{u_{x32}+u_{x21}}{1+\frac{u_{x32}u_{x21}}{c^2}} \qquad \text{may be simplified as:} \qquad u_3 = \frac{u_1+u_2}{1+\frac{u_1 u_2}{c^2}}$$

where u_1 is the velocity of O_1 relative to O, u_2 is the velocity of O_2 relative to O_1, and u_3 is the velocity of O_2 relative to O. What are the advantages and disadvantages of each form?
Answer: The second form is obviously simpler. However, the first form contains more information and it is more systematic and rigorous since the x subscript indicates that all the motions are occurring along the same orientation and hence it excludes the cases where the motions are in two different directions, i.e. x and y or x and z. Moreover, each double-numeric subscript in the first form (i.e. 31, 32 and 21) indicates exactly which object/observer the velocity belongs to and which object/observer the velocity is relative to, e.g. 31 means the velocity of 3 relative to 1. We therefore prefer to use the first form.

4.2.3 Velocity Composition

13. Compose two equal velocities to obtain a composite velocity that is $0.8c$ where all these velocities have the same orientation.[79]

    Answer: Since all the velocities have the same orientation then the appropriate formula to use is the x velocity composition formula where the composite velocity is $u_{x31} = 0.8c$ while the composed velocities (assumed to be in the positive direction) are $u_{x21} = u_{x32} \equiv u$, that is:

    $$u_{x31} = \frac{u_{x32} + u_{x21}}{1 + \frac{u_{x32} u_{x21}}{c^2}}$$

    $$u_{x31} = \frac{u + u}{1 + \frac{u^2}{c^2}}$$

    $$u_{x31} + u_{x31}\frac{u^2}{c^2} = 2u$$

    $$0.8c\frac{u^2}{c^2} - 2u + 0.8c = 0$$

    $$0.8u^2 - 2cu + 0.8c^2 = 0$$

    This can be solved by the quadratic formula to obtain:

    $$u = \frac{2c \pm \sqrt{4c^2 - 4 \times 0.8 \times 0.8c^2}}{2 \times 0.8}$$

    so $u = 0.5c$ or $u = 2c$. Although both solutions are acceptable mathematically, the second solution should be rejected because it is non-physical according to the formalism of Lorentz mechanics where no velocity can exceed c (at least for massive objects) and hence the solution is $u = 0.5c$.[80] If the composite velocity in the question was $-0.8c$ then we similarly define $u_{x21} = u_{x32} \equiv u$ (where u is understood to be negative since the composite velocity is in the negative direction) and hence we have:

    $$0.8u^2 + 2cu + 0.8c^2 = 0 \quad \text{and} \quad u = \frac{-2c \pm \sqrt{4c^2 - 4 \times 0.8 \times 0.8c^2}}{2 \times 0.8}$$

    so we obtain $u = -0.5c$ or $u = -2c$ where the accepted solution is $u = -0.5c$. So, in both cases we obtain the same composed speed. We note that if the composed velocities have opposite directions (i.e. they have the same magnitude but opposite signs) then the composite velocity will be zero (which should be obvious) and that is why in the question we stated "equal velocities" not "equal speeds".

14. Using the velocity composition formula, show that the composition of two velocities which are less than c is less than c. Assume that both velocities are in the positive x direction.

---

[79] In questions like this (which may be seen as ambiguous) it should be obvious that the orientation is determined by a primary inertial observer. Also, "equal velocities" should be sufficient to remove any ambiguity.

[80] In fact, this is also based on common sense since the composed speeds should not exceed the composite speed where all these speeds are in the same direction.

Answer: Let have $u_{x32} = ac$ and $u_{x21} = bc$ where $0 \leq a, b < 1$. Using the x velocity composition formula we have:

$$u_{x31} = \frac{u_{x32} + u_{x21}}{1 + \frac{u_{x32}u_{x21}}{c^2}} = \frac{ac + bc}{1 + ab} = \frac{a+b}{1+ab}c$$

Now, for $0 \leq a, b < 1$ we have:

$$\frac{a+b}{1+ab} < 1$$

and hence $u_{x31} < c$.

4.2.4 Acceleration Transformations

1. O and O' are two inertial observers in a state of standard setting with relative velocity $v = 0.3c$. Observer O measures the velocity and acceleration of an object at a given instant of time to be $0.2c$ and 6 in the x direction and $0.1c$ and 4 in the z direction. What is the acceleration of this object at that time according to observer O' in these directions?
   Answer: Using the Lorentz acceleration transformations from O frame to O' frame, we have: $v = 0.3c$, $u_x = 0.2c$, $a_x = 6$, $u_z = 0.1c$ and $a_z = 4$. Hence:

$$a'_x = \frac{a_x}{\gamma^3 \left(1 - \frac{vu_x}{c^2}\right)^3} = \frac{6\left(1 - 0.3^2\right)^{3/2}}{(1 - 0.3 \times 0.2)^3} \simeq 6.2709 \, \text{m/s}^2$$

$$a'_z = \frac{c^2 a_z - v u_x a_z + v u_z a_x}{c^2 \gamma^2 \left(1 - \frac{vu_x}{c^2}\right)^3}$$

$$= \frac{[4c^2 - (0.3c \times 0.2c \times 4) + (0.3c \times 0.1c \times 6)](1 - 0.3^2)}{c^2 \times (1 - 0.3 \times 0.2)^3} \simeq 4.3167 \, \text{m/s}^2$$

2. As stated in the text, while in classical mechanics the acceleration is value-invariant across inertial frames, in Lorentz mechanics the acceleration is frame dependent. Does this mean that the inertiality status (i.e. being inertial or non-inertial) of a frame could be different between these mechanics, i.e. a frame can be inertial in one of these mechanics but non-inertial in the other mechanics? If not, show that according to Lorentz mechanics if the acceleration vanishes in an inertial frame, it should vanish in all inertial frames.[81]
   Answer: No, and hence a frame is either inertial in classical and Lorentz mechanics or non-inertial in both mechanics. The reason is that if the acceleration vanishes in a frame, then it will vanish in any other frame according to the definition of both mechanics. Hence, if we start from a known inertial frame (say the absolute frame), then the inertiality status of all other frames will be the same in both mechanics because both mechanics will agree that the transformed acceleration is either zero or not although

---

[81] The acceleration in this question belongs to a frame of reference rather than an object.

they may disagree on the value of acceleration if it is not zero.

More technically, from the given Lorentz acceleration transformations we see that if the acceleration in O/O' frame vanishes (i.e. $a_x = a_y = a_z = 0$ or $a'_x = a'_y = a'_z = 0$) then it will also vanish in O'/O frame. Now, since O and O' are arbitrary inertial frames, then we can conclude that if the acceleration vanishes in an inertial frame it will vanish in all inertial frames (and if not it will not). This may be used to conclude that the inertiality status of a frame is the same in both classical mechanics and Lorentz mechanics although the transformation rules of acceleration (and hence the quantitative values) in these mechanics are different. So in brief, all inertial frames are the same in both mechanics because the two mechanics agree on vanishing acceleration, and similarly all non-inertial frames are the same in both mechanics because the two mechanics agree on non-vanishing acceleration although the two mechanics may disagree on the value of the non-vanishing acceleration of a given non-inertial frame.

3. What other conclusion can be drawn from the argument in the previous question?
   Answer: Since all inertial frames agree on vanishing acceleration of physical objects, then in Lorentz mechanics, like in classical mechanics, Newton's first law is invariant across all inertial frames. In fact, the validity of this argument also requires the presumption of the objectivity of force and hence if force is observed (or not observed) in a given frame then it should be observed (or not observed) in all other frames.

4.2.5 Length Transformation

4.2.6 Time Interval Transformation

4.2.7 Mass Transformation

1. Discuss mass transformation in Lorentz mechanics.
   Answer: According to the old view, the mass of a massive object is frame dependent and hence it is an extrinsic property although the proper mass is an intrinsic property. So, according to the old view, the mass is transformed between its rest frame and a moving frame by the formula:
   $$m' = \gamma m$$
   while according to the modern view the mass is an invariant property, and hence we have:
   $$m' = m$$
   We note that the symbols m and m' belong to the rest frame and moving frame.

4.2.8 Frequency Transformation and Doppler Shift

1. A light source of frequency $\nu_0 = 6.6 \times 10^{14}$ is located at the origin of coordinates of an inertial frame S. An inertial observer O is receding from the light source with speed $v = 10^8$. What is the frequency ν and the wavelength λ of the signal as observed by O?

Answer: Since O is receding from the source then we have:

$$\nu = \nu_0 \sqrt{\frac{c-v}{c+v}} \simeq 6.6 \times 10^{14} \times \sqrt{\frac{3 \times 10^8 - 10^8}{3 \times 10^8 + 10^8}} \simeq 4.6669 \times 10^{14}\,\text{Hz}$$

$$\lambda = \frac{c}{\nu} \simeq 642.82\,\text{nm}$$

i.e. the signal is red shifted as it should be.

2. An electromagnetic transmitter of wavelength $\lambda_0 = 10^{-8}$ is located at the origin of coordinates of an inertial frame S. An inertial observer O observed that the transmitted signal has a constant wavelength $\lambda = 9 \times 10^{-9}$. What are the speed v and the angle θ of the relative motion between the observer and the transmitter?

Answer: Since $\lambda < \lambda_0$ then the signal is obviously blue shifted. Moreover, since the observed wavelength λ is constant and the frames are inertial then the observer should be heading towards the transmitter with a constant speed, and hence we should have:

$$\nu = \nu_0 \sqrt{\frac{c+v}{c-v}}$$

Accordingly:

$$\lambda = \frac{c}{\nu}$$

$$\lambda = \frac{c}{\nu_0} \sqrt{\frac{c-v}{c+v}}$$

$$\lambda = \lambda_0 \sqrt{\frac{c-v}{c+v}}$$

$$(\lambda/\lambda_0)^2 = \frac{c-v}{c+v}$$

$$(\lambda/\lambda_0)^2 c + (\lambda/\lambda_0)^2 v = c - v$$

$$v + (\lambda/\lambda_0)^2 v = c - (\lambda/\lambda_0)^2 c$$

$$v = \frac{1 - (\lambda/\lambda_0)^2}{1 + (\lambda/\lambda_0)^2} c$$

$$v = \frac{1 - 0.9^2}{1 + 0.9^2} c$$

$$v \simeq 0.1050 c$$

Moreover, since the observer is heading towards the transmitter then we have:

$$\theta = 0$$

3. An electromagnetic transmitter of wavelength $\lambda_0 = 10^{-7}$ is located at the origin of coordinates of an inertial frame S. An observer O observed that the transmitted signal has a constant wavelength $\lambda = 1.1 \times 10^{-7}$. What are the speed v and the angle θ of

4.2.8 Frequency Transformation and Doppler Shift

the relative motion between the observer and the transmitter? Assume that the speed of the observer is constant and $\theta \ne \pi$.

Answer: Since $\lambda > \lambda_0$ then the signal is obviously red shifted. Moreover, since the observed wavelength λ is constant and $\theta \ne \pi$ then O should be moving transversely with a constant speed v around the source following a circular path that is centered at the source and hence O is not an inertial observer.[82] If we assume that the stated formulae in the text equally apply to the instantaneous rest frame (see § 1.3) of a non-inertial observer like O then we should have:

$$\nu = \nu_0 \sqrt{1 - \beta^2}$$

Accordingly:

$$\begin{aligned}
\lambda &= \frac{c}{\nu} \\
\lambda &= \frac{c}{\nu_0 \sqrt{1 - \beta^2}} \\
\sqrt{1 - \beta^2} &= \lambda_0 / \lambda \\
\beta^2 &= 1 - (\lambda_0/\lambda)^2 \\
\beta &= \sqrt{1 - (\lambda_0/\lambda)^2} \\
v &= c\sqrt{1 - (1/1.1)^2} \simeq 0.4166c
\end{aligned}$$

Moreover, since the observer is following a circular path that is centered at the source then we should have:

$$\theta = \pi/2$$

4. The sodium D_2 line of a star is observed from the Earth to be red shifted by 11 nm. What is the velocity of the star relative to the Earth?

Answer: The sodium D_2 line has a proper wavelength $\lambda_0 \simeq 589$ nm. Hence, the observed wavelength is $\lambda \simeq 589 + 11 \simeq 600$ nm. Also, because the line is red shifted then the star is receding from the Earth. Accordingly, we have:

$$\begin{aligned}
\nu &= \nu_0 \sqrt{\frac{c - v}{c + v}} \\
\frac{c}{\lambda} &= \frac{c}{\lambda_0} \sqrt{\frac{c - v}{c + v}} \\
\lambda &= \lambda_0 \sqrt{\frac{c + v}{c - v}} \\
(\lambda/\lambda_0)^2 (c - v) &= c + v
\end{aligned}$$

---

[82] In fact, this possibility (i.e. constant speed around the source following a circular path) is the simplest case but it is not the only one, and that is why we assumed in the question that the speed of the observer is constant. Other possibilities of variable translational or rotational acceleration should also be possible. The reader may consider these possibilities as other exercises.

4.2.8 Frequency Transformation and Doppler Shift

$$(\lambda/\lambda_0)^2 v + v = (\lambda/\lambda_0)^2 c - c$$
$$v = \frac{(\lambda/\lambda_0)^2 - 1}{(\lambda/\lambda_0)^2 + 1} c$$
$$v \simeq \frac{(600/589)^2 - 1}{(600/589)^2 + 1} c$$
$$v \simeq 0.0185 c$$

So, the star is receding from the Earth along the line of sight with a speed of about $0.0185c$.

5. A radio transmitter of frequency $\nu_0 = 10^8$ is located at the origin of coordinates of an inertial frame S. An inertial observer O whose speed in S is $v = 0.6c$ measured the frequency at $t = 0$ to be $\nu = 8 \times 10^7$. What is the velocity of O relative to the transmitter at $t = 0$?

Answer: The speed of O is given and hence it is known. So, to determine the velocity we need to know the direction of motion of O relative to the transmitter at $t = 0$, and this direction is determined by the angle θ. Accordingly, we have:

$$\nu = \frac{\nu_0}{\gamma \left(1 - \frac{v}{c} \cos\theta\right)}$$
$$1 - \frac{v}{c} \cos\theta = \frac{\nu_0}{\nu \gamma}$$
$$\cos\theta = \frac{c}{v}\left(1 - \frac{\nu_0}{\nu \gamma}\right)$$
$$\cos\theta = \frac{1}{0.6}\left(1 - \frac{10^8}{8 \times 10^7} \times \sqrt{1 - 0.6^2}\right)$$
$$\cos\theta = \frac{1}{0.6}\left(1 - \frac{1}{0.8} \times 0.8\right)$$
$$\cos\theta = 0$$
$$\theta = \pi/2$$

Hence, at $t = 0$ the observer O is moving transversely relative to the transmitter with speed $v = 0.6c$.[83]

6. An observer at the origin of coordinates of an inertial frame O measured the frequency of a radio signal coming from an inertial spaceship S_1 to be $\nu_1 = 10^9$. If S_1 is receding from the observer along the positive x direction with a speed $v_1 = 0.5c$ relative to O, (a) What is the proper frequency ν_0 of the radio signal? (b) What is the frequency ν_2 as measured by a second inertial spaceship S_2 that recedes from the observer along the negative x direction with a speed $v_2 = 0.4c$ relative to O?

Answer:

---

[83] In fact, this is not a complete determination of velocity, but it should be enough for our purpose which may be guessed from "the velocity of O relative to the transmitter".

4.2.8 Frequency Transformation and Doppler Shift

(a) Because S_1 is receding from O, the proper frequency ν_0 of the radio signal should be obtained from the formula:
$$\nu = \nu_0 \sqrt{\frac{c-v}{c+v}}$$
that is:
$$\nu_0 = \nu_1 \sqrt{\frac{c+v_1}{c-v_1}} = 10^9 \times \sqrt{\frac{c+0.5c}{c-0.5c}} = 10^9 \sqrt{3}\,\text{Hz}$$

(b) To find ν_2 we need to know the relative speed between S_1 and S_2 by using the Lorentzian velocity composition formula for the x coordinate. Now, if we label S_2, O and S_1 with 1, 2 and 3 then we have $u_{x32} = 0.5c$, $u_{x21} = 0.4c$ and hence the required speed is u_{x31}, that is:
$$u_{x31} = \frac{u_{x32} + u_{x21}}{1 + \frac{u_{x32} u_{x21}}{c^2}} = \frac{0.5c + 0.4c}{1 + (0.5 \times 0.4)} = 0.75c$$

Hence, the frequency ν_2 is given by:
$$\nu_2 = \nu_0 \sqrt{\frac{c - u_{x31}}{c + u_{x31}}} = 10^9 \sqrt{3} \times \sqrt{\frac{c - 0.75c}{c + 0.75c}} = 10^9 \sqrt{3/7}\,\text{Hz}$$

7. The equation of classical Doppler frequency shift is given by:
$$\nu = \frac{c + v_r}{c + v_s} \nu_0$$

where ν and ν_0 are the received (improper) and emitted (proper) frequency, while v_r and v_s are the velocity of the receiver and source relative to the medium of propagation along the line of sight. The sign of v_r is positive/negative when the receiver is approaching/receding from the source, while the sign of v_s is positive/negative when the source is receding/approaching the receiver. Show that the Lorentzian Doppler frequency converges to this classical limit when (a) The receiver is stationary with respect to the medium and the source is receding from the receiver. (b) The source is stationary with respect to the medium and the receiver is receding from the source.

Answer:
(a) If the receiver is stationary relative to the medium then $v_r = 0$ and hence the classical formula becomes:
$$\nu = \frac{c}{c + v_s} \nu_0$$
The Lorentzian formula for this case (ignoring the medium) is given by:
$$\nu = \nu_0 \sqrt{\frac{c-v}{c+v}} = \nu_0 \frac{\sqrt{c^2 - v^2}}{c+v} = \frac{c\sqrt{1 - (v/c)^2}}{c+v} \nu_0$$

where in the second equality we multiplied the numerator and denominator by $\sqrt{c+v}$. Now, in the classical limit $v \ll c$, $(v/c)^2$ becomes negligible and hence the Lorentzian

formula becomes identical to the classical formula when $v_s = v$ noting that v_s is positive according to the sign convention that we stated in the question.
(b) If the source is stationary relative to the medium then $v_s = 0$ and hence the classical formula becomes:
$$\nu = \frac{c + v_r}{c}\nu_0$$
The Lorentzian formula for this case (ignoring the medium) is given by:
$$\nu = \nu_0 \sqrt{\frac{c-v}{c+v}} = \nu_0 \frac{c-v}{\sqrt{c^2 - v^2}} = \frac{c-v}{c\sqrt{1-(v/c)^2}}\nu_0$$
where in the second equality we multiplied the numerator and denominator by $\sqrt{c-v}$. Now, in the classical limit $v \ll c$, $(v/c)^2$ becomes negligible and hence the Lorentzian formula becomes identical to the classical formula when $v_r = -v$ noting that v_r is negative according to the sign convention that we stated in the question.

4.2.9 Charge Density and Current Density Transformations

1. A charged object whose charge density in its rest frame is $\rho_0 = 6$ is seen from an inertial frame O_1 to have a charge density $\rho_1 = 10$. What is the charge density ρ_2 and the current density $\mathbf{j}_2$ as seen from another inertial frame O_2 which is moving uniformly relative to O_1 with a velocity of $0.2c$? Assume that the motion of O_1 and O_2 relative to the object is in the positive x direction.
Answer: We label O_1, O_2 and the rest frame of the charged object with 1, 2 and 3. Accordingly, u_{x31} is the velocity of the object relative to O_1 and hence:
$$\rho_1 = \gamma \rho_0$$
$$\rho_1 = \frac{\rho_0}{\sqrt{1 - (|u_{x31}|/c)^2}}$$
$$1 - (|u_{x31}|/c)^2 = (\rho_0/\rho_1)^2$$
$$|u_{x31}| = c\sqrt{1 - (\rho_0/\rho_1)^2}$$
$$|u_{x31}| = c\sqrt{1 - 0.6^2}$$
$$|u_{x31}| = 0.8c$$

Now, since O_1 is assumed to move in the positive x direction relative to the object, then $u_{x31} = -0.8c$. Moreover, $u_{x21} = 0.2c$. So, to find the velocity u_{x32} of the object relative to O_2, we use the Lorentzian velocity composition formula for the x direction, that is:
$$u_{x32} = \frac{u_{x31} - u_{x21}}{1 - \frac{u_{x31}u_{x21}}{c^2}} = \frac{-0.8c - 0.2c}{1 - \frac{(-0.8c)(0.2c)}{c^2}} = \frac{-1}{1.16}c \simeq -0.8621c$$

Therefore:

$$\rho_2 = \frac{\rho_0}{\sqrt{1-(u_{x32}/c)^2}} \simeq \frac{6}{\sqrt{1-(-0.8621)^2}} \simeq 11.8392\,\mathrm{C/m^3}$$

$$\mathbf{j}_2 = \frac{\rho_0 \mathbf{u}}{\sqrt{1-(u_{x32}/c)^2}} = \frac{6 \times (-0.8621c, 0, 0)}{\sqrt{1-(-0.8621)^2}} \simeq \left(-3.0619 \times 10^9, 0, 0\right)\,\mathrm{C/(m^2 .s)}$$

4.3 Physical Relations

4.3.1 Newton's Second Law

1. Show that all Newton's laws of motion are valid in Lorentz mechanics. What you conclude?
   Answer: Newton's second law in its Lorentzian form, as given in the text, is obviously valid in Lorentz mechanics. Newton's first law is no more than a special case of Newton's second law and hence it should also be valid in Lorentz mechanics (also refer to the exercises of § 4.1.7).[84] The essence of Newton's third law is coupling of forces and hence if Newton's second law (which may be regarded as a definition of force) is valid in Lorentz mechanics and Newton's third law is valid in classical mechanics then Newton's third law should also be valid in Lorentz mechanics if we use the Lorentzian form of Newton's second law to define force.[85]
   Conclusion: all the laws of classical mechanics, which are based on Newton's laws of motion, should also be valid in Lorentz mechanics if we use the Lorentzian forms and definitions of the involved quantities.
2. Compare Newton's second law in classical mechanics and in Lorentz mechanics.
   Answer: In classical mechanics, we have two forms of Newton's second law. The first is "force equals time derivative of momentum" and the second is "force equals mass times acceleration". Lorentz mechanics agrees with classical mechanics on the first form (considering the Lorentzian definition of momentum) but not on the second form since the Lorentzian force is given by $f = ma\gamma^3$ and hence, unlike classical mechanics, force is not proportional to acceleration.

4.3.2 Mass-Energy Relation

1. It is common in some branches of physics to use energy units as mass units (with proper conversion factors) and vice versa. Why?
   Answer: This is mainly for convenience because the conversion between mass and energy units becomes easier. For example, if we express a given mass in units of MeV/c^2 then we can obtain the equivalent energy of this mass easily by just dropping

---

[84] In this regard, we refer to the following Lorentzian form of Newton's second law: $f = ma\gamma^3$ where force vanishes *iff* acceleration is zero noting that $m \neq 0$ and $\gamma \neq 0$.
[85] The reader is referred to the literature about the conditions for the validity of Newton's third law.

4.3.2 Mass-Energy Relation

c^2 from the denominator of the units and keeping the numerical value. So, knowing that the mass of electron is 0.511 MeV/c^2, we immediately know that its rest energy is 0.511 MeV with no need for calculations to convert mass to energy. This advantage is a dominant factor in some branches of physics, such as particle physics, where the mass-energy equivalence relation is in common use and hence the conversion is frequently needed.

2. Find the equivalent energy of the mass of neutron (i.e. its rest energy) giving your answer in SI units.
   Answer: We have:
   $$E_{0n} = m_n c^2 \simeq 1.6749 \times 10^{-27} \,\text{kg} \times \left(3 \times 10^8 \,\text{m/s}\right)^2 \simeq 1.5074 \times 10^{-10} \,\text{J}$$
   where E_{0n} is the rest energy (i.e. E_0) of neutron and m_n is its mass.

3. Find the energy required to combine an electron with a proton to produce a neutron.
   Answer: We have $m_e \simeq 0.511$, $m_p \simeq 938.27$ and $m_n \simeq 939.57$ where all these masses are in units of MeV/c^2. Accordingly, the difference in mass, which represents the binding energy, is given by:
   $$\Delta m = m_n - (m_e + m_p) \simeq 0.789 \,\text{MeV}/c^2$$
   Hence, according to the Poincare mass-energy relation, the required binding energy (which may also be called the reaction energy according to the convention that we mentioned in the text) is:
   $$\Delta E = \Delta m c^2 \simeq 0.789 \,(\text{MeV}/c^2) \times c^2 = 0.789 \,\text{MeV} \simeq 1.2641 \times 10^{-13} \,\text{J}$$
   We can also work in standard SI units from the start where $m_e \simeq 9.1094 \times 10^{-31}$ kg, $m_p \simeq 1.6726 \times 10^{-27}$ kg and $m_n \simeq 1.6749 \times 10^{-27}$ kg to get a similar final answer.

4. A parent particle with mass m_P decays, while being at rest, into two daughter particles: A with mass m_A and B with mass m_B. If the velocity of A is $u_A = -0.4c$ and the velocity of B is $u_B = 0.6c$ (where both velocities are along a single orientation), what is the mass of the parent particle in terms of m_A and m_B?
   Answer: From the Poincare mass-energy relation plus the principle of conservation of total energy before and after the decay, we have:
   $$\begin{aligned} m_P c^2 &= m_A c^2 \gamma_A + m_B c^2 \gamma_B \\ m_P &= m_A \gamma_A + m_B \gamma_B \\ m_P &= \frac{m_A}{\sqrt{1 - 0.4^2}} + \frac{m_B}{\sqrt{1 - 0.6^2}} \\ m_P &\simeq 1.0911 m_A + 1.25 m_B \end{aligned}$$

   If we follow the old Lorentzian convention about mass, then the problem can be solved by using the conservation of mass (which we may call the Lorentzian mass) before and after decay where we start from the second line in the above sequence of equations.

4.3.2 Mass-Energy Relation

5. The temperature of a block of aluminum of mass $m_b = 10$ kg was increased by $\Delta T = 1000$ K. What is the relative increase in mass?
   Answer: The specific heat capacity of aluminum is $C \simeq 921.2$ J/(kg.K) and hence the added heat is:[86]

$$E_h = m_b C \Delta T \simeq 10 \times 921.2 \times 1000 = 9.212 \times 10^6 \text{ J}$$

This energy is equivalent to a mass of about:

$$m_h = \frac{E_h}{c^2} \simeq 1.0236 \times 10^{-10} \text{ kg}$$

Hence, the relative mass increase Δm_r is about:

$$\Delta m_r \simeq \frac{m_h}{m_b} \simeq 1.0236 \times 10^{-11}$$

6. The total energy of a particle of mass m is to be increased to 1.8 of its rest value. What is the required speed to realize this?
   Answer: We should have:

$$\begin{aligned} E_t &= 1.8 E_0 \\ \gamma m c^2 &= 1.8 m c^2 \\ \gamma &= 1.8 \\ \sqrt{1-\beta^2} &= 5/9 \\ 1-\beta^2 &= 25/81 \\ \beta^2 &= 56/81 \\ u &= c\sqrt{56}/9 \end{aligned}$$

So, the required speed is about $0.8315c$.

7. Object A of mass m_A and speed $u_A = 0.9c$ makes a perfectly inelastic collision with object B of mass $m_B = 2m_A$ which is at rest. What is the mass m_t of the single object that results from this collision in terms of m_A?
   Answer: If we label the speed of the resulting single object with u, then from the conservation of total energy before and after collision plus mass-energy equivalence we have:

$$\frac{m_t c^2}{\sqrt{1-u^2/c^2}} = \frac{m_A c^2}{\sqrt{1-0.9^2}} + 2 m_A c^2$$

$$\frac{m_t}{\sqrt{1-u^2/c^2}} = \frac{m_A}{\sqrt{1-0.9^2}} + 2 m_A$$

---

[86] We are assuming that the specific heat capacity is independent of temperature; otherwise an integral is needed, i.e. $E_h = \int m_b C \, dT$ where this integral is evaluated between the two temperature limits.

4.3.2 Mass-Energy Relation

Also, from the conservation of momentum before and after collision we have:

$$\frac{m_t u}{\sqrt{1 - u^2/c^2}} = \frac{m_A 0.9c}{\sqrt{1 - 0.9^2}}$$

On substituting from the second line of the energy equation into the momentum equation and canceling m_A we obtain:

$$\left(\frac{m_A}{\sqrt{1 - 0.9^2}} + 2m_A\right) u = \frac{m_A 0.9c}{\sqrt{1 - 0.9^2}}$$

$$u = \left(\frac{0.9c}{\sqrt{1 - 0.9^2}}\right) \left(\frac{1}{\sqrt{1 - 0.9^2}} + 2\right)^{-1}$$

$$u \simeq 0.4808c$$

And hence from the second line of the energy equation we obtain:

$$m_t = \sqrt{1 - u^2/c^2} \left(\frac{m_A}{\sqrt{1 - 0.9^2}} + 2m_A\right) \simeq 3.7652 m_A$$

8. A radioactive nucleus of mass 109 amu (atomic mass unit) decays and releases $E_r = 7.4724 \times 10^{-11}$ J of energy. What is the total mass that is left after the decay?
   Answer: The mass of 1 amu is $m_{\text{amu}} \simeq 1.6605 \times 10^{-27}$ kg. Hence, the equivalent mass of the released energy is:

$$\Delta m = \frac{E_r}{c^2} \simeq 8.3027 \times 10^{-28} \text{ kg} \simeq 0.5 \text{ amu}$$

So, the total mass that is left after the decay is about 108.5 amu.

9. A proton at rest is accelerated through a potential difference $\Delta V = 10^7$ V. What is its speed?
   Answer: The charge of proton is identical in magnitude to the charge of electron. Hence, the kinetic energy of the proton E_{kp} is:

$$E_{kp} = e\Delta V = 1 \times 10^7 = 10^7 \text{ eV} = 10 \text{ MeV}$$

The rest energy of proton is $E_{0p} \simeq 938.27$ MeV. Hence:

$$E_{kp} = m_p c^2 (\gamma - 1)$$
$$E_{kp} = E_{0p} (\gamma - 1)$$
$$10 = 938.27 (\gamma - 1)$$
$$\gamma = \frac{10}{938.27} + 1$$
$$\sqrt{1 - \beta^2} = \left(\frac{10}{938.27} + 1\right)^{-1}$$
$$1 - \beta^2 = \left(\frac{10}{938.27} + 1\right)^{-2}$$

$$\beta^2 = 1 - \left(\frac{10}{938.27} + 1\right)^{-2}$$

$$\beta = \sqrt{1 - \left(\frac{10}{938.27} + 1\right)^{-2}}$$

$$u \simeq 0.1448c$$

4.3.3 Momentum-Energy Relation

1. Some may argue that since we have $E = \gamma mc^2$ and $p = \gamma mu$, then from the relation $E = pc$ for a massless particle we have:

$$E = pc$$
$$\gamma mc^2 = \gamma muc$$
$$c = u$$

i.e. the speed of massless particles must be c. Assess this argument.
Answer: This is not a valid argument because for massless particles $m = 0$ and hence both sides of the second equality should vanish which makes the conclusion of the third equality questionable. In fact, the use of the formula $E = pc$ which is for massless objects already assumes that the mass is zero and hence it is meaningless to use the equations $E = \gamma mc^2$ and $p = \gamma mu$ which involve mass and hence they are specific to massive objects.

2. Assess the following argument which may be found in some textbooks: for a massless object, such as a photon, that is supposed to move with a speed lower than c, the expressions for total energy (i.e. $E = \gamma mc^2$) and momentum (i.e. $p = \gamma mu$) will vanish because $m = 0$, i.e. $E = p = 0$. Now, since this is untrue because E and p are not zero, then this should indicate that no massless object can move with a speed lower than c to avoid this contradiction.
Answer: This argument is not valid. Again, there is a contradiction between having a massless object and using the formulae $E = \gamma mc^2$ and $p = \gamma mu$ which are specific to massive objects. Accordingly, the premise that "no massless object can move with a speed lower than c" should be established by a more reliable argument.

3. Let assume that according to the formalism of Lorentz mechanics, no massive object can reach or exceed c and a massless object must not move with a speed less than c. What about massless objects moving with speeds greater than c? Is there a relation in Lorentz mechanics that implies the impossibility of exceeding c for massless objects?
Answer: It seems that there is nothing in the current formalism of Lorentz mechanics that prevents this case, i.e. massless objects moving at speeds higher than c. For example, if we assume the validity[87] of the velocity transformation (and composition) formulae for cases in which some velocities exceed c then the composite velocity (as well

---

[87] This assumption is argumentative because the Lorentz spacetime coordinate transformations (from which the velocity transformations are derived) are limited by certain speed restrictions which can be

4.3.3 Momentum-Energy Relation

as the composed velocity) may exceed c in some cases, e.g. if we use the u_{x31} formula with $u_{x21} = 0.1c$ and $u_{x32} = 1.5c$ then we have:

$$u_{x31} = \frac{u_{x32} + u_{x21}}{1 + \frac{u_{x32} u_{x21}}{c^2}} = \frac{1.5c + 0.1c}{1 + 1.5 \times 0.1} \simeq 1.39c$$

which does not seem inconsistent. Accordingly, even if we accept (based on the formalism or something else) the propositions that the speed of massive objects cannot reach or exceed c and the speed of massless objects cannot be lower than c, we still need a valid argument to establish the proposition that the speed of massless objects cannot exceed c. We should remark that the argument against this case by the fact that γ becomes imaginary when $u > c$ is not valid since we should also reject (for a similar reason) the case $u = c$ for this type of objects because γ becomes singular when $u = c$. More discussion about this issue will follow.

4. The total energy of an electron is 0.53 MeV. What is the magnitude of its momentum?
   Answer: The rest energy of electron is about 0.511 MeV. Hence, we have:

$$\begin{aligned} E^2 &= p^2 c^2 + m^2 c^4 \\ p^2 c^2 &= E^2 - E_0^2 \\ p^2 c^2 &\simeq (0.53\,\text{MeV})^2 - (0.511\,\text{MeV})^2 \\ p^2 c^2 &\simeq 0.0198\,\text{MeV}^2 \\ pc &\simeq 0.1406\,\text{MeV} \\ p &\simeq 0.1406\,\text{MeV}/c \end{aligned}$$

5. What is the total energy of a neutron whose momentum is $p = 50\,\text{MeV}/c$?
   Answer: The rest energy of neutron is $E_0 \simeq 939.57$ MeV. We use the momentum-energy formula, that is:

$$\begin{aligned} E^2 &= p^2 c^2 + m_n^2 c^4 \\ E &= \sqrt{p^2 c^2 + E_0^2} \\ E &= \sqrt{(50\,\text{MeV}/c)^2 \times c^2 + (939.57\,\text{MeV})^2} \\ E &= \sqrt{(50\,\text{MeV})^2 + (939.57\,\text{MeV})^2} \\ E &\simeq 940.90\,\text{MeV} \\ E &\simeq 1.5075 \times 10^{-10}\,\text{J} \end{aligned}$$

6. Find the kinetic energy of a proton whose momentum is 150 MeV/c.
   Answer: The rest energy of proton is $E_0 \simeq 938.27$ MeV. We use the momentum-energy

---

claimed to include even the massless objects. In fact, according to this the assumption that some of the composed velocities exceed c can be rejected in the first place. So, the purpose of this example and its alike is to explore possibilities more than to establish arguments.

4.3.3 Momentum-Energy Relation

relation, that is:

$$\begin{aligned} E^2 &= p^2c^2 + m_p^2c^4 \\ (E_0 + E_k)^2 &= p^2c^2 + E_0^2 \\ E_0^2 + 2E_0E_k + E_k^2 &= p^2c^2 + E_0^2 \\ 2E_0E_k + E_k^2 &= p^2c^2 \\ E_k^2 + 2E_0E_k - p^2c^2 &= 0 \\ E_k^2 + 1876.54E_k - 150^2 &= 0 \end{aligned}$$

Using the quadratic formula, we obtain:

$$E_k \simeq \frac{-1876.54 \pm \sqrt{1876.54^2 - 4 \times 1 \times (-150^2)}}{2 \times 1}$$

that is: $E_k \simeq -1888.45\,\text{MeV}$ or $E_k \simeq 11.91\,\text{MeV}$. The negative root should be rejected because it is non-physical.

This problem may also be solved by using the definition of momentum (i.e. $p = \gamma mu$) plus the definition of kinetic energy, i.e. $E_k = E_0\,(\gamma - 1)$. To ease the notation we use the following symbol:

$$A \equiv \frac{p}{cm_p} \simeq \frac{150\,\text{MeV}/c}{c \times 938.27\,\text{MeV}/c^2} = \frac{150}{938.27}$$

Hence:

$$\begin{aligned} p &= \frac{m_p u}{\sqrt{1 - \beta^2}} \\ \frac{\beta}{\sqrt{1 - \beta^2}} &= \frac{p}{cm_p} \\ \frac{\beta^2}{1 - \beta^2} &= A^2 \\ \beta^2 &= \frac{A^2}{1 + A^2} \\ \sqrt{1 - \beta^2} &= \sqrt{1 - \frac{A^2}{1 + A^2}} \\ \gamma &= \left(1 - \frac{A^2}{1 + A^2}\right)^{-1/2} \end{aligned}$$

Therefore:

$$E_k = E_0\,(\gamma - 1) = 938.27 \times \left[\left(1 - \frac{A^2}{1 + A^2}\right)^{-1/2} - 1\right] \simeq 11.91\,\text{MeV}$$

7. What is the advantage of expressing the energy in terms of momentum rather than velocity?
   Answer: Because the momentum is conserved but the velocity is not.

4.3.4 Work-Energy Relation

1. A force is given by: $\mathbf{f} = (x, y^2, \sqrt{z})$. What is the work done by this force in moving a particle from position $\mathbf{r}_1 = (1.2, -2.7, 5.3)$ to position $\mathbf{r}_2 = (-3.4, 2\pi, 3.9)$? What is the change in the kinetic energy of the particle in moving between these positions?
Answer: From the definition of work we have:

$$
\begin{aligned}
W &= \int_{\mathbf{r}_1}^{\mathbf{r}_2} \mathbf{f} \cdot d\mathbf{r} \\
&= \int_{\mathbf{r}_1}^{\mathbf{r}_2} \left(x, y^2, \sqrt{z}\right) \cdot (dx, dy, dz) \\
&= \int_{1.2}^{-3.4} x\,dx + \int_{-2.7}^{2\pi} y^2\,dy + \int_{5.3}^{3.9} \sqrt{z}\,dz \\
&= \left[\frac{x^2}{2}\right]_{1.2}^{-3.4} + \left[\frac{y^3}{3}\right]_{-2.7}^{2\pi} + \left[\frac{2}{3}z^{3/2}\right]_{5.3}^{3.9} \\
&\simeq 5.06 + 89.2444 - 2.9998 \\
&\simeq 91.3046\,\text{J}
\end{aligned}
$$

From the work-energy relation we have:

$$\Delta E_k = W \simeq 91.3046\,\text{J}$$

4.4 Conservation Laws

1. Discuss the conservation laws of mass and energy in classical and Lorentz mechanics.
Answer: In classical mechanics, mass and energy are conserved as two independent physical quantities. But in Lorentz mechanics, the two are conserved as a single quantity because of the equivalence between mass and energy according to the Poincare relation and hence what is conserved is the combined mass-energy which may be seen as conservation of Lorentzian mass $m = \gamma m_0$ (where all forms of non-kinetic energy are included in the rest mass) or conservation of total energy $E_t = \gamma m_0 c^2$. In more details, we have the following cases of conservation:
• Conservation of mass where there is no conversion between mass and energy.
• Conservation of energy where there is no conversion between mass and energy or between various forms of energy.
• Conservation of energy where there is no conversion between mass and energy but there is a conversion from one form of energy to another form of energy with the total energy (in its classical sense) being conserved.
• Conservation of mass-energy where there is a potential conversion between mass and energy and a potential conversion between different forms of energy.[88]

---

[88] The case of conservation of "mass and energy" rather than "mass-energy" (assuming that the system contains both) where there is no conversion between mass and energy (with or without conversion between different forms of energy) can be represented as a combination of the first three cases.

4.4 Conservation Laws

We note that these cases of conservation equally apply in classical and Lorentz mechanics except the last case which applies only in Lorentz mechanics since there is no conversion between mass and energy in classical mechanics. Accordingly, the laws of classical mechanics of conservation of mass and conservation of energy can be considered as special cases for the more general law of conservation of mass-energy in Lorentz mechanics (i.e. by excluding any potential conversion between mass and energy in the last case).

However, we should remark that the equivalence between mass and energy, as expressed by the Poincare relation, may be established by purely classical arguments (refer to § 5.3.2) and hence the conservation of mass-energy (instead of conservation of mass and conservation of energy) may not be proprietary to Lorentz mechanics since the classical framework can provide sufficient theoretical justification for this equivalence. If this is the case, then both mechanics have the same law of mass-energy conservation (although this law was not considered in the early history of classical mechanics) and hence they agree in all the aforementioned four cases of conservation.

2. A bound electron captured a 2 MeV photon. Neglecting the recoil of the atom to which the electron is bound and assuming that the electron is initially at rest, find the speed of the electron after capture assuming that the external forces on the electron are negligible.
Answer: If we label the energy of the photon with E_{ph}, the rest energy of electron with E_{0e} and the total energy of the electron after capture with E_t, then from the conservation of total energy we have:

$$E_t = m_e c^2 + E_{\text{ph}}$$
$$\gamma m_e c^2 = m_e c^2 + E_{\text{ph}}$$
$$\gamma E_{0e} = E_{0e} + E_{\text{ph}}$$
$$\gamma = \frac{E_{0e} + E_{\text{ph}}}{E_{0e}}$$
$$\frac{1}{\sqrt{1-\beta^2}} \simeq \frac{0.511 + 2}{0.511}$$
$$1 - \beta^2 = \left(\frac{0.511}{2.511}\right)^2$$
$$\beta = \sqrt{1 - \left(\frac{0.511}{2.511}\right)^2}$$
$$u \simeq 0.9791c$$

3. A particle A of mass m and kinetic energy $E_{kA} = 2mc^2$ collides with a stationary particle B of mass $1.5m$ and they coalesce. What are the mass, energy and momentum of the composite particle in terms of m and c assuming that there is no loss of mass or energy to the environment in this process?
Answer: If we symbolize the speed of particle A with u then from the definition of

kinetic energy we have:

$$\begin{aligned} E_{kA} &= E_{0A}(\gamma - 1) \\ \gamma &= \frac{E_{kA}}{E_{0A}} + 1 \\ \gamma &= \frac{2mc^2}{mc^2} + 1 \\ \gamma &= 3 \\ u &\simeq 0.9428c \end{aligned}$$

Now, if we label the mass and speed of the composite particle with M and v then from the conservation of momentum we have:

$$\frac{mu}{\sqrt{1-(u/c)^2}} = \frac{Mv}{\sqrt{1-(v/c)^2}}$$

and from the conservation of total energy we have:

$$\frac{mc^2}{\sqrt{1-(u/c)^2}} + 1.5mc^2 = \frac{Mc^2}{\sqrt{1-(v/c)^2}}$$

From the latter equation we obtain:

$$M = \sqrt{1-(v/c)^2}\left(\frac{m}{\sqrt{1-(u/c)^2}} + 1.5m\right)$$

On substituting this into the momentum equation we obtain:

$$\frac{mu}{\sqrt{1-(u/c)^2}} = \left(\frac{m}{\sqrt{1-(u/c)^2}} + 1.5m\right)v$$

$$v = \frac{u}{\sqrt{1-(u/c)^2}}\left(\frac{1}{\sqrt{1-(u/c)^2}} + 1.5\right)^{-1}$$

$$v \simeq 0.6285c$$

Therefore, the mass, energy and momentum of the composite particle are:

$$M = \sqrt{1-(v/c)^2}\left(\frac{m}{\sqrt{1-(u/c)^2}} + 1.5m\right) = 3.5m$$

$$E = \frac{Mc^2}{\sqrt{1-(v/c)^2}} = 4.5mc^2$$

$$p = \frac{Mv}{\sqrt{1-(v/c)^2}} \simeq 2.8284mc$$

4.5 Restoring Classical Formulation at Low Speed

1. Give examples (other than those given in the text) for the convergence of the Lorentzian formulation to the classical formulation at the low-speed limit, i.e. $v \ll c$.
   Answer: Other examples include:
   • Transformations of space coordinates and time where the Lorentzian transformations reduce to their Galilean form at the limit of low speed:[89]
   $$x' = \gamma(x - vt) \quad \rightarrow \quad x' \simeq x - vt \quad (\gamma \rightarrow 1)$$
   $$t' = \gamma\left(t - \beta\frac{x}{c}\right) \quad \rightarrow \quad t' \simeq t \quad (\beta \rightarrow 0, \gamma \rightarrow 1)$$
   • Force:
   $$f = ma\gamma^3 \quad \rightarrow \quad f \simeq ma \quad (\gamma \rightarrow 1)$$

2. Discuss the conditions for the validity of the classical formula of velocity transformation in the x direction. Also discuss the validity of the approximation in the y and z direction.
   Answer: The Lorentzian formula for the velocity transformation in the x direction is given by:
   $$u_x = \frac{u'_x + v}{1 + \frac{vu'_x}{c^2}}$$
   As we see, there are two velocities on the right hand side, i.e. u'_x and v. In the text we considered only one of these velocities, i.e. v. However, u'_x, like v, should also be subject to the same low speed restriction since its role in the formula is similar to the role of v as seen in the above equation. So, the condition for the applicability of the classical form is that the denominator of the Lorentzian form approaches 1, i.e. $\left(1 + \frac{vu'_x}{c^2}\right) \simeq 1$ or $|vu'_x| \ll c^2$. The situation will be more clear if we use the formula of velocity composition, which is another form of the velocity transformation formula as shown in § 4.2.3, that is:
   $$u_{x31} = \frac{u_{x32} + u_{x21}}{1 + \frac{u_{x32}u_{x21}}{c^2}}$$
   where the symmetry of the formula in u_{x32} and u_{x21} is more obvious.
   The velocity transformation in the y direction, i.e.
   $$u_y = \frac{u'_y}{\gamma\left(1 + \frac{vu'_x}{c^2}\right)}$$
   also converges to the classical form (i.e. $u_y \simeq u'_y$) when v and u'_x are very small compared to c and hence $\gamma \simeq 1$ and $\left(1 + \frac{vu'_x}{c^2}\right) \simeq 1$, i.e. the denominator approaches 1. This also applies to the velocity transformation in the z direction since it is identical in form to the velocity transformation in the y direction.

---

[89] We note that for the time transformation another condition should be imposed on x, however this is not related to speed which is the matter of concern in this section.

3. Find the speed condition for which the relative difference between the Lorentzian and classical velocity transformation relations for the x dimension does not exceed 0.05. Comment on the question.
Answer: The classical and Lorentzian velocity transformation relations for the x dimension are given by:

$$u'_{xc} = u_x - v \quad \text{and} \quad u'_{xl} = \frac{u_x - v}{1 - \frac{vu_x}{c^2}}$$

Hence:

$$u'_{xl} = \frac{u_x - v}{1 - \frac{vu_x}{c^2}}$$

$$u'_{xl} = \frac{u'_{xc}}{1 - \frac{vu_x}{c^2}}$$

$$\frac{u'_{xl}}{u'_{xc}} = \frac{1}{1 - \frac{vu_x}{c^2}}$$

$$\frac{u'_{xc}}{u'_{xl}} = 1 - \frac{vu_x}{c^2}$$

$$1 - \frac{u'_{xc}}{u'_{xl}} = \frac{vu_x}{c^2}$$

$$\left|\frac{u'_{xl} - u'_{xc}}{u'_{xl}}\right| = \left|\frac{vu_x}{c^2}\right|$$

So, the speed condition is: $|vu_x| \leq 0.05c^2$.
Comment: this example provides a general method for setting the required limit where classical mechanics is a good approximation to Lorentz mechanics within a given allowed error margin.

4.6 Restrictions at High Speed

1. Give some examples for the speed restrictions on massive objects that are based on the current formalism of Lorentz mechanics.
   Answer: There are several examples; however we give only two obvious examples:
   • Momentum: when a massive object approaches the characteristic speed of light, the Lorentzian expression for momentum (i.e. $p = \gamma mu$) tends to infinity.
   • Energy: when a massive object approaches the characteristic speed of light, the Lorentzian expression for energy (i.e. $E = \gamma mc^2$) tends to infinity.
   These examples, as well as other similar examples, may be seen as an indication that no massive object can reach the speed of light. By priority, no massive object can exceed the speed of light within the domain of Lorentz mechanics and according to its current formulation.
2. Discuss the issue of the speed of physical objects within the framework of Lorentz mechanics and beyond.

4.6 Restrictions at High Speed

Answer: Regarding the speed of physical objects, there are two main cases: massive objects and massless objects:

• Within the domain of validity of Lorentz mechanics and within the current formulation, the speed of massive objects should not reach or exceed c, as can be inferred for example from the Lorentzian formulae for momentum and energy. However, this restriction may be lifted outside the validity domain or/and by future extension or modification to the current formulation.

• The speed of massless objects can certainly reach c as it is the case with light. However, should these objects strictly have this speed, i.e. is it impossible for a massless object to have a speed lower or higher than c? There seems to be no requirement of the current formulation of Lorentz mechanics and within its domain of validity for this restriction.[90] Hence, further evidence is required to establish the generally accepted proposition that the speed of all massless objects should be c. Again, even if the case for this restriction is established within the framework of Lorentz mechanics, it is still possible for this restriction to be lifted outside the domain of validity of Lorentz mechanics or/and by a future extension or modification to the current formulation.[91]

3. An electron is seen from an inertial frame O_1 to have a momentum $p_1 = 2\,\text{MeV}/c$ and from another inertial frame O_2 to have a momentum $p_2 = 4\,\text{MeV}/c$. What is the relative speed between O_1 and O_2 assuming that O_1 and O_2 are in a state of standard setting and the momenta are referred to their common x axis? Comment on the result.

Answer: We label $\frac{p_1 c}{m_e c^2}$ as A and hence we have:

$$A \equiv \frac{p_1 c}{m_e c^2} = \frac{p_1 c}{E_{0e}} \simeq \frac{(2\,\text{MeV}/c) \times c}{0.511\,\text{MeV}} \simeq 3.914$$

From O_1 we have:

$$p_1 = \gamma_1 m_e u_1$$
$$\gamma_1 u_1 = \frac{p_1}{m_e}$$
$$\frac{u_1}{\sqrt{1-\beta_1^2}} = \frac{p_1}{m_e}$$
$$\frac{\beta_1}{\sqrt{1-\beta_1^2}} = \frac{p_1 c}{m_e c^2}$$
$$\frac{\beta_1}{\sqrt{1-\beta_1^2}} = A$$
$$\frac{\beta_1^2}{1-\beta_1^2} = A^2$$

---

[90] We note that the case of light should be excluded due to the invariance of its speed according to the current formulation and within the validity domain of Lorentz mechanics. As discussed before and will be investigated further, this invariance is based on the use of the speed of light in spacetime calibration and hence it does not necessarily extend to all massless objects.

[91] This lift could include even the speed of light as discussed earlier.

4.6 Restrictions at High Speed

$$\beta_1^2 = \frac{A^2}{1+A^2}$$
$$u_1 \simeq 0.9689c$$

where the negative root is discarded because p_1 is positive. Similarly, from O_2 we have:

$$u_2 \simeq 0.9919c$$

Now, since the momenta are referred to the x axis we use the velocity composition formula for the x dimension where we label O_1, O_2 and electron with 1, 2 and 3. Hence, $u_{x31} \equiv u_1 \simeq 0.9689c$, $u_{x32} \equiv u_2 \simeq 0.9919c$, and the required speed is $|u_{x21}|$, that is:

$$u_{x21} = \frac{u_{x31} - u_{x32}}{1 - \frac{u_{x31}u_{x32}}{c^2}} \simeq \frac{0.9689c - 0.9919c}{1 - (0.9689 \times 0.9919)} \simeq -0.5923c$$

and hence the relative speed between O_1 and O_2 is $|u_{x21}| \simeq 0.5923c$.

Comment: the Lorentzian velocity composition formulae may be seen to provide another example for the restriction on the speeds in Lorentz mechanics because as we saw in this example some of the composite relative speeds should exceed c if we follow the Galilean additive rule of velocity composition. However, the reader is referred to the next question about the nature of this restriction.

4. Can we infer from the velocity composition and velocity transformation formulae that the speed of massless objects is restricted to c (i.e. the speed cannot be less than or greater than c)? What about the restriction on the speed of massive objects to be less than c?

   Answer: The velocity composition and velocity transformation formulae mean that the velocities are transformed between frames in this manner where there is a limit on the observed composite velocity to be less than or equal to c when the composed velocities are less than or equal to c. This does not mean that falling below c or exceeding c by massless objects is impossible in its own. In fact, having a speed less than c by massless objects cannot be ruled out because of these formulae since there is no restriction on this case from these formulae. Yes, for exceeding c there may be a legitimate argument for the restriction since these formulae produce results that are restricted to be less than or equal to c but this applies when the composed velocities are less than or equal to c. As shown earlier in an example (see § 4.3.3), if we assume that some of the composed velocities exceed c then these formulae legitimately yield a composite velocity that exceeds c. In brief, these formulae put a restriction on the composite velocity to be less than or equal to c when the composed velocities are assumed to be less than or equal to c, but these formulae do not put a restriction on the composed velocities to be less than or equal to c and hence when the composed velocities are assumed to exceed c the formulae yield sensible results where the composite velocity could also exceed c.[92]

---

[92] The point of this is not to assert that the composition formulae apply to the cases where the composed velocities exceed c and hence the composite velocity also exceeds c, but to reject the claim that these formulae impose a limit on the velocity. Yes, as indicated before (see § 4.3.3), the composition formulae are derived from the Lorentz spacetime coordinate transformations which are limited by certain speed restrictions that can be claimed to include even massless objects of all types. Accordingly, the restriction will ultimately rest on this tentative claim.

We should remark that the above deliberation is based on an implicit assumption that these formulae equally apply to all types of massless objects as to massive objects, and this assumption may be debated.

With regard to massive objects, the speed restriction on massive objects to be less than c can be legitimately inferred from other Lorentzian formulae like the formula of momentum or the formula of energy. Hence, the velocity composition and velocity transformation formulae are not needed to establish this restriction regardless of being eligible to impose a speed restriction or not.

Chapter 5
Derivation of Formalism

1. Can we consider the limitation that is imposed in Lorentz mechanics on the physical speeds (i.e. $v < c$) as evidence for the proposal that c is a restricted speed to light and it is an ultimate speed for all physical objects?
   Answer: The Lorentzian limitation on physical speeds, i.e. $v < c$, does not imply that it is a universal fact. The reason is that the theory of Lorentz mechanics could be a limited theory representing a limiting case for a more general theory, as it is the case with classical mechanics which is a limiting case to Lorentz mechanics or the claim that Lorentz mechanics (as commonly represented by special relativity) is a limiting case to general relativity. So, although the formalism of Lorentz mechanics is restricted to $v < c$, this does not represent an evidence that $v < c$ is a universal fact. In fact, even within the domain of applicability of Lorentz mechanics (i.e. inertial frames) we can imagine that the condition $v < c$ may not apply because the current formulation of this mechanics is not complete. For example, we can imagine that the Lorentz γ factor includes terms in β of higher order which are negligible in most experimental situations. We should remark that because the domain of validity of Lorentz mechanics is restricted to inertial frames, the restriction $v < c$ does not extend automatically to non-inertial frames even if the premise about c as being restricted and ultimate speed is accepted as a universal fact within the domain and formalism of Lorentz mechanics. We should also point out to the fact which was indicated earlier that the case of massless objects should be treated differently from the case of massive objects with regard to speed. So, even if the restriction $v < c$ is finally established on the speeds of massive objects by the Lorentzian formalism or by any other evidence, we still need an independent evidence for the restrictions that are commonly imposed within the framework of special relativity on the speed of massless objects[93] where it is claimed that this speed is strictly restricted to c and hence it cannot fall below or exceed c.

5.1 Physical Transformations

1. Why we prefer to postulate the Lorentz spacetime coordinate transformations and derive the other transformations and formulations of Lorentz mechanics from these postulates instead of deriving the Lorentz spacetime coordinate transformations from other postulates as it is the case in special relativity where they are derived from the postulate of relativity and the postulate of constancy of speed of light in all inertial frames?
   Answer: The reason is that: basing the formulation of Lorentz mechanics on the Lorentz spacetime coordinate transformations as postulates makes this mechanics more objective and neutral to epistemological and philosophical interpretations and hence it

---
[93] As seen earlier, we should exclude light due to the invariance of its speed according to the current formalism of Lorentz mechanics.

5.1.1 Lorentz Spacetime Coordinate Transformations

will be open to various interpretations. On the other hand, basing the formulation of this mechanics on the postulates of special relativity or any other theory will bind this mechanics to the epistemological and philosophical framework of this theory and force the investigator to adopt views that may not be supported by scientific evidence.

5.1.1 Lorentz Spacetime Coordinate Transformations

1. What is the physical significance of the Lorentz spacetime coordinate transformations and to which physical phenomena they are related?
   Answer: Our belief is that the Lorentz spacetime coordinate transformations originate from the effects of length contraction and time dilation which are experienced under the influence of motion. As we will see in the forthcoming parts of the book, this connection can be demonstrated and presented in various forms with various justifications (refer for example to § 5.1.4, § 5.1.5 and § 12.4.2).
2. Discuss the procedural aspects in solving the problems related to the Lorentz spacetime coordinate transformations.
   Answer: In brief, the Lorentz transformations have nine variables or parameters: four spacetime coordinates for each frame (i.e. $x, y, z, t, x', y', z', t'$) plus the relative velocity (i.e. v assuming a state of standard setting) between the two frames. So, given sufficient information any problem can be solved either directly or by manipulating the given transformations or/and combining more than one transformation. Many of these methods and techniques have been demonstrated in § 4.2.1. Sometimes the space or time separation (or interval) between two events is required; in such cases the difference between the corresponding spacetime coordinates of the events should be considered, as demonstrated in the text and as will be seen in the next exercise.
3. Write the expressions for the space and time intervals between two events V_A and V_B as measured by O' knowing the spacetime coordinates of the events as measured by O where O and O' are in a state of standard setting.
   Answer:
   $$\begin{aligned}
   x'_B - x'_A &= \gamma(x_B - vt_B) - \gamma(x_A - vt_A) = \gamma\left[(x_B - x_A) - v(t_B - t_A)\right] \\
   y'_B - y'_A &= y_B - y_A \\
   z'_B - z'_A &= z_B - z_A \\
   t'_B - t'_A &= \gamma\left(t_B - \frac{vx_B}{c^2}\right) - \gamma\left(t_A - \frac{vx_A}{c^2}\right) = \gamma\left[(t_B - t_A) - \frac{v}{c^2}(x_B - x_A)\right]
   \end{aligned}$$
4. Discuss if the mingling of space and time into spacetime is a complete novelty of Lorentz mechanics.
   Answer: We can consider the intertwining of the space coordinates and time in the above Lorentz transformations as a distinctive feature of the Lorentz transformations and Lorentz mechanics and hence this mix of space and time into spacetime may be seen as something novel that has been introduced by Lorentz mechanics. However, we should also note that the Galilean transformation of x coordinate (i.e. $x' = x - vt$)

also contains some sort of intertwining. In fact, the Galilean and Lorentzian transformations of the x spatial coordinate are identical apart from the Lorentz γ factor which does not include a spatial or temporal coordinate and it is dimensionless. So, what is entirely novel in the Lorentz transformations and Lorentz mechanics is the mixing in the time transformation, as can be seen by comparing the equation $t' = t$ of the Galilean transformation of time to the equation $t' = \gamma \left(t - \frac{vx}{c^2}\right)$ of the Lorentz transformation of time. So, the mingling of space and time in Lorentz mechanics is obviously more extensive and fundamental and has certain novelty aspects (and this is reflected for example in the invariance of spacetime interval and non-invariance of space interval and time interval) although it is not entirely novel.[94]

5.1.2 Velocity Transformations

1. Derive the Lorentz velocity transformations from the unprimed frame to the primed frame using this time the differential method instead of the chain rule of differentiation.
   Answer: Using the Lorentz transformations for x' and t' (i.e. $x' = \gamma(x - vt)$ and $t' = \gamma\left(t - \frac{vx}{c^2}\right)$), we have:

   $$dx' = \gamma(dx - vdt)$$
   $$dt' = \gamma\left(dt - \frac{vdx}{c^2}\right)$$

   Hence:

   $$u'_x = \frac{dx'}{dt'} = \frac{\gamma(dx - vdt)}{\gamma\left(dt - \frac{vdx}{c^2}\right)} = \frac{\frac{dx}{dt} - v}{1 - \frac{v}{c^2}\frac{dx}{dt}} = \frac{u_x - v}{1 - \frac{vu_x}{c^2}}$$

   where we divided the numerator and denominator by dt in the third step and used the definition of velocity in the last step. Similarly, we have:

   $$u'_y = \frac{dy'}{dt'} = \frac{dy}{\gamma\left(dt - \frac{vdx}{c^2}\right)} = \frac{\frac{dy}{dt}}{\gamma\left(1 - \frac{v}{c^2}\frac{dx}{dt}\right)} = \frac{u_y}{\gamma\left(1 - \frac{vu_x}{c^2}\right)}$$

   $$u'_z = \frac{dz'}{dt'} = \frac{dz}{\gamma\left(dt - \frac{vdx}{c^2}\right)} = \frac{\frac{dz}{dt}}{\gamma\left(1 - \frac{v}{c^2}\frac{dx}{dt}\right)} = \frac{u_z}{\gamma\left(1 - \frac{vu_x}{c^2}\right)}$$

5.1.3 Acceleration Transformations

1. Derive the Lorentz acceleration transformations from the unprimed frame to the primed frame using this time the differential method instead of the chain rule of differentiation.
   Answer: Using the transformation equations for u'_x and t' (i.e. $u'_x = \frac{u_x - v}{1 - \frac{vu_x}{c^2}}$ and

---

[94] We should also refer to the effect of the speed of relative motion (as represented by the γ factor) on the spacetime coordinates which is a total novelty of Lorentz mechanics.

5.1.3 Acceleration Transformations

$t' = \gamma\left(t - \frac{vx}{c^2}\right)$), we have:

$$du'_x = \frac{du_x\left(1 - \frac{vu_x}{c^2}\right) - (u_x - v)\left(-\frac{vdu_x}{c^2}\right)}{\left(1 - \frac{vu_x}{c^2}\right)^2} = \frac{du_x\left(1 - \frac{v^2}{c^2}\right)}{\left(1 - \frac{vu_x}{c^2}\right)^2} = \frac{du_x}{\gamma^2\left(1 - \frac{vu_x}{c^2}\right)^2}$$

$$dt' = \gamma\left(dt - \frac{vdx}{c^2}\right)$$

Hence:

$$\begin{aligned}
a'_x &= \frac{du'_x}{dt'} \\
&= \frac{du_x}{\gamma^2\left(1 - \frac{vu_x}{c^2}\right)^2 \gamma\left(dt - \frac{vdx}{c^2}\right)} \\
&= \frac{a_x}{\gamma^3\left(1 - \frac{vu_x}{c^2}\right)^2\left(1 - \frac{vu_x}{c^2}\right)} \\
&= \frac{a_x}{\gamma^3\left(1 - \frac{vu_x}{c^2}\right)^3}
\end{aligned}$$

where the third step is obtained by dividing the numerator and denominator by dt and using the definitions of acceleration and velocity.

Similarly, from the transformation equation for u'_y (i.e. $u'_y = \frac{u_y}{\gamma\left(1 - \frac{vu_x}{c^2}\right)}$) we have:

$$du'_y = \frac{du_y\gamma\left(1 - \frac{vu_x}{c^2}\right) - u_y\gamma\left(-\frac{vdu_x}{c^2}\right)}{\gamma^2\left(1 - \frac{vu_x}{c^2}\right)^2} = \frac{du_y - du_y\frac{vu_x}{c^2} + u_y\frac{vdu_x}{c^2}}{\gamma\left(1 - \frac{vu_x}{c^2}\right)^2}$$

Hence:

$$\begin{aligned}
a'_y &= \frac{du'_y}{dt'} \\
&= \frac{du_y - du_y\frac{vu_x}{c^2} + u_y\frac{vdu_x}{c^2}}{\gamma\left(1 - \frac{vu_x}{c^2}\right)^2 \gamma\left(dt - \frac{vdx}{c^2}\right)} \\
&= \frac{a_y - a_y\frac{vu_x}{c^2} + u_y\frac{va_x}{c^2}}{\gamma^2\left(1 - \frac{vu_x}{c^2}\right)^2\left(1 - \frac{vu_x}{c^2}\right)} \\
&= \frac{a_y - a_y\frac{vu_x}{c^2} + u_y\frac{va_x}{c^2}}{\gamma^2\left(1 - \frac{vu_x}{c^2}\right)^3} \\
&= \frac{c^2 a_y - vu_x a_y + vu_y a_x}{c^2\gamma^2\left(1 - \frac{vu_x}{c^2}\right)^3}
\end{aligned}$$

where the third step is obtained by dividing the numerator and denominator by dt and using the definitions of acceleration and velocity. The derivation of a'_z is identical to the derivation of a'_y.

5.1.4 Length Transformation

1. Find a general formula that correlates the total length of an object in one inertial frame O to its total length in another inertial frame O' where O and O' are in a state of standard setting and the object is arbitrarily oriented in any direction.
Answer: We have:

$$L = |\mathbf{L}| = \sqrt{L_x^2 + L_y^2 + L_z^2} = \sqrt{\gamma^2 (L_x')^2 + (L_y')^2 + (L_z')^2}$$

where L is the length of the object in frame O while (L_x, L_y, L_z) and (L_x', L_y', L_z') are its components in frames O and O'.

5.1.5 Time Interval Transformation

1. Use a single argument to derive both the Lorentzian length transformation and the Lorentzian time interval transformation. Comment on the results.
Answer: Let have an inertial observer O with an object of length L in his rest frame and another inertial observer O' that moves at speed v relative to O along the orientation of the object. According to O, O' passes through L in time (of O'):

$$\Delta t' = \frac{L}{v} \quad \Rightarrow \quad L = v\Delta t'$$

while according to O' the object has length L' and he passes through this length in time (of O):

$$\Delta t = \frac{L'}{v} \quad \Rightarrow \quad L' = v\Delta t$$

Now, if we assume only the time dilation effect, then we have:

$$\frac{L}{L'} = \frac{v\Delta t'}{v\Delta t} = \frac{v\gamma\Delta t}{v\Delta t} = \gamma \quad \Rightarrow \quad L = \gamma L' \qquad (1)$$

which is the length contraction formula. Similarly, if we assume only the length contraction effect, then we have:

$$\frac{\Delta t}{\Delta t'} = \frac{v\Delta t}{v\Delta t'} = \frac{L'}{L} = \frac{L'}{\gamma L'} = \frac{1}{\gamma} \quad \Rightarrow \quad \Delta t' = \gamma \Delta t \qquad (2)$$

which is the time dilation formula.
Comment: there are two things to note:
• This derivation shows that time dilation and length contraction have a common origin, i.e. there is a single fundamental physical principle (i.e. the contraction of spacetime coordinates with motion) behind both of these effects and that is why we can find a single argument from which both effects can be obtained.
• This confirms our explanation about the difference in perspective when we use the terms "length contraction" and "time dilation" (instead of "length contraction" and "time

5.1.5 Time Interval Transformation 167

contraction" or "length dilation" and "time dilation") and that is why we write $L = v\Delta t'$ and $L' = v\Delta t$ (instead of $L = v\Delta t$ and $L' = v\Delta t'$).
The suspected reader who may think this argument is wrong should analyze the meaning of the used symbols carefully first, and should explain why we should use $L = x_2 - x_1$ (which is based on using the transformation of space coordinates from the primed to the unprimed) to derive length contraction and $\Delta t' = t'_2 - t'_1$ (which is based on using the transformation of time coordinates from the unprimed to the primed) to derive time dilation. He is also advised to read the answer to the next exercise.

2. Repeat the previous question to show "time contraction" and "length dilation". Also, comment on the results.
Answer: Let have an inertial observer O with an object of length L in his rest frame and another inertial observer O' that moves at speed v relative to O. According to O, O' passes through L in time (of O):

$$\Delta t = \frac{L}{v} \quad \Rightarrow \quad L = v\Delta t$$

while according to O' the object has length L' and he passes through this length in time (of O'):

$$\Delta t' = \frac{L'}{v} \quad \Rightarrow \quad L' = v\Delta t'$$

Now, if we assume only the time dilation effect, then we have:

$$\frac{L}{L'} = \frac{v\Delta t}{v\Delta t'} = \frac{v\Delta t}{v\gamma\Delta t} = \frac{1}{\gamma} \quad \Rightarrow \quad L' = \gamma L \qquad (3)$$

which is the "length dilation" formula. Similarly, if we assume only the length contraction effect, then we have:

$$\frac{\Delta t}{\Delta t'} = \frac{v\Delta t}{v\Delta t'} = \frac{L}{L'} = \frac{\gamma L'}{L'} = \gamma \quad \Rightarrow \quad \Delta t = \gamma\Delta t' \qquad (4)$$

which is the "time contraction" formula.
Comment: this reveals that the labels of the two effects are based on the above mentioned difference in perspective. Hence, when we unified our perspectives we obtained time dilation and length dilation in one case and length contraction and time contraction in the other case. Accordingly, length contraction and time dilation effects represent in essence the same effect that is experienced by the spatial and temporal coordinates of spacetime where these coordinates are seen to contract (or dilate as seen from an opposite perspective) by the effect of motion.

3. Considering the procedural aspect of time measurement, what distinguishes proper time interval from improper time interval?
Answer: Proper time means the time as measured by a clock which is at rest relative to the observer while improper time means the time as measured by a clock which is in motion relative to the observer. Now, because proper time interval is measured at rest, the start and end of the proper time interval are measured by the same clock

(or equivalently at the same location) in the frame to which this interval belongs. On the other hand, because improper time interval is measured in a state of motion, the start and end of the improper time interval are measured by two different clocks in the frame to which this interval belongs. To put it in another way, since the proper time interval represents the time as measured at the same location by a given observer, a single clock is needed to measure the interval. In contrast, since the improper time interval represents the time as measured at two different locations by a given observer, two different (but synchronized) clocks are needed to measure the interval.

4. Provide more clarifications about the conditions that should be satisfied to justify using time dilation and length contraction formulae instead of the Lorentz spacetime transformations.
   Answer: For time dilation relation to apply, the two events should be co-positional to one observer. Similarly, for length contraction relation to apply, the two events should be simultaneous to one observer. So, if the events are not co-positional to any observer then the Lorentz time transformation should be used to find the temporal separation between the two events. Similarly, if the events are not simultaneous to any observer then the Lorentz space transformation should be used to find the spatial separation between the two events. This means that when time dilation effect applies then one observer needs a single clock while the other observer needs two synchronized clocks, so if both observers need two synchronized clocks (i.e. each in his own frame) to record the time of the events then the Lorentz time transformation should be used.

5.1.6 Mass Transformation

1. Clarify the issue of the invariance of mass according to the old and modern conventions of Lorentz mechanics.
   Answer: According to the modern convention "mass" is mass (i.e. it is the same whether at rest or in motion) and hence the invariance of mass and the invariance of rest mass are the same and they both hold according to this convention. However, according to the old convention the rest mass is different from the non-rest mass and hence while the invariance of rest mass holds the invariance of mass in general does not hold. In other words, while the invariance of rest mass is true in both conventions, the invariance of non-rest mass (and hence the invariance of mass in general) is true only in the modern convention.

2. Following the old convention about mass, provide an argument based on the conservation of momentum to show that the mass is transformed from its rest frame to a moving frame according to $m = \gamma m_0$.
   Answer: Let have two observers, O and O', who are in a state of standard setting with relative speed v. Observer O fires a bullet in the y direction towards a block of wood which is in his rest frame and the bullet penetrates the block creating a hole of length L. Now, since O and O' agree on the lengths in the y direction which is perpendicular to the direction of their relative motion they should agree on L. It is sensible to assume that the length of the hole is solely dependent on the momentum of the bullet in the

y direction. Accordingly, since O and O' agree on the length of the hole they should agree on the momentum of the bullet in the y direction. Now, if we use the classical definition of momentum then we should have:

$$p'_y = m'u'_y = mu_y = p_y$$

On transforming the y component of the velocity from O' to O noting that $u_x = 0$ we obtain (with γ being a function of v):

$$\begin{aligned} m'u'_y &= mu_y \\ m'\frac{u_y}{\gamma} &= mu_y \\ m'u_y &= \gamma mu_y \\ m' &= \gamma m \end{aligned}$$

which is the required result with m and m' being the rest and non-rest mass (i.e. corresponding to m_0 and m in the relation $m = \gamma m_0$ which is given in the question).[95] We note that this "proof" is not sufficiently rigorous since it is based on several assumptions and approximations (e.g. using the classical definition of momentum and assuming that the γ factor belongs exclusively to mass) some of which may not be obvious. We should also note that this type of argument may not fit well within a certain theoretical structure due for example to circularity; so the purpose of bringing this sort of questions and answers is diversity rather than building a rigorous Lorentzian theory.

5.1.7 Frequency Transformation and Doppler Shift

5.1.8 Charge Density and Current Density Transformations

1. Provide more clarification about the transformation of electric current density.
   Answer: To be more clear, we have: $\mathbf{j}' = \rho_0 \mathbf{0}$ in frame O'. By transforming the two sides of this equation from O' to O, we obtain: $\mathbf{j} = \gamma \rho_0 \mathbf{u}$ which is a combination of the transformations: $\mathbf{j}' \to \mathbf{j}$, $\rho_0 \to \gamma \rho_0$ and $\mathbf{0} \to \mathbf{u}$.

5.2 Physical Quantities

5.2.1 Momentum

1. Make a simple argument to obtain the Lorentzian form of momentum with no need for formal derivation.
   Answer: Although the following argument is not rigorous, it should be useful for pedagogical purposes. Let follow the old formalism of Lorentz mechanics where mass is considered as a frame dependent quantity and hence we have $m' = m\gamma$ where m and

---

[95] The "rest" and "non-rest" here is with respect to the x orientation which is the orientation of the relative motion between O and O'.

5.2.1 Momentum

m' are the proper and improper mass of the object. The classical form of momentum (i.e. $p = m'u$)[96] will then apply even in Lorentz mechanics, that is:

$$p = m'u = \gamma m u$$

As indicated earlier, this definition satisfies all the aforementioned four requirements of a legitimate Lorentzian definition of momentum, and hence it is a legitimate form.

2. Make a simple argument (based on the validity of conservation of momentum in classical mechanics) to justify the third requirement of an acceptable Lorentzian definition of momentum.
   Answer: The third requirement can be simply established by the validity of the conservation of momentum in classical mechanics plus the fact that the Lorentzian formulation should converge to the classical formulation in the classical limit (which the first requirement is based upon), and hence we need a Lorentzian principle that conserves momentum in its classical domain and in the extended Lorentzian domain and this obviously requires a suitable definition of momentum that satisfies this demand. Accordingly, the third and first requirements may be seen to have a common basis.

3. Make a simple argument (based on the validity of Newton's second law in classical mechanics) to justify the fourth requirement of an acceptable Lorentzian definition of momentum.
   Answer: The argument should be very similar to the argument of the previous question. In brief, the necessity for the convergence of Lorentzian formulation to the classical formulation plus the validity of Newton's second law in the classical domain requires a momentum formulation (or "definition") that keeps the essential relation between force and momentum (and even the relation between force and acceleration considering the convergence) as expressed by the Newton's second law. The alternative is to create a totally new formulation that replaces Newton's second law even in its classical domain and this requires a fundamental change in physics which (assuming that such a change is possible) is not an attractive choice. Accordingly, Newton's second law is kept as in classical mechanics while the definition of momentum is modified to adapt with the Newton's second law both in the classical domain and in the extended Lorentzian domain. Therefore, the fourth and first requirements may be seen to have a common basis (and hence if we except the second requirement then all the requirements have a common basis which is the necessity for the convergence of Lorentz mechanics to classical mechanics at the classical limit).

4. What is the implication of the following statement: "The components of momentum in the perpendicular directions to the direction of relative motion between the frames of observation are value invariant across these frames and hence only the component in the direction of relative motion is variant".
   Answer: The implication is that the direction of momentum is not invariant across frames (where the direction is defined in terms of a common spatial coordinate system or by the relative size of the components).

---

[96] For convenience, we use here m' instead of m where this is justified by the fact that $m = m'$ according to classical mechanics.

5.2.2 Force

1. Write down the mathematical form of Newton's second law in Lorentz mechanics.
   Answer: The 1D version of the Lorentzian form of Newton's second law is:
   $$f = ma\gamma^3$$
   where f is force, m is mass, a is acceleration, and γ is the Lorentz factor as a function of the velocity u of the object and where we are assuming that f, a and u are in the same direction in this 1D version. The above form is based on the constancy of mass (i.e. no exchange of mass between the object and its surrounding). The more general form of Newton's second law is given by:
   $$f = \frac{dp}{dt} = \frac{d}{dt}(\gamma m u)$$
   where p is momentum and t is time.

5.2.3 Energy

1. Evaluate the following integral analytically using a method other than the integration by parts which is used in the text (see § 1.4):
   $$\int_{x_1}^{x_2} \frac{d}{dt}(\gamma m u)\, dx$$

   Answer:
   $$\begin{aligned}
   \int_{x_1}^{x_2} \frac{d}{dt}(\gamma m u)\, dx &= m \int_{x_1}^{x_2} \frac{d}{dt}(\gamma u)\, dx \\
   &= m \int_{x_1}^{x_2} \left[u \frac{d\gamma}{dt} dx + \gamma \frac{du}{dt} dx \right] \\
   &= m \int_{t_1}^{t_2} \left[u \frac{d\gamma}{dt} \frac{dx}{dt} + \gamma \frac{du}{dt} \frac{dx}{dt} \right] dt \\
   &= m \int_{t_1}^{t_2} \left[u^2 \frac{d\gamma}{dt} + \gamma u \frac{du}{dt} \right] dt \\
   &= m \int_{t_1}^{t_2} \left[c^2 \beta^2 \frac{d\gamma}{dt} + \gamma c^2 \beta \frac{d\beta}{dt} \right] dt \\
   &= mc^2 \int_{t_1}^{t_2} \left[\beta^2 \frac{d\gamma}{dt} + \gamma \beta \frac{d\beta}{dt} \right] dt \\
   &= mc^2 \int_{t_1}^{t_2} \left[\beta^2 \frac{\beta \frac{d\beta}{dt}}{(1-\beta^2)^{3/2}} + \frac{\beta \frac{d\beta}{dt}}{(1-\beta^2)^{1/2}} \right] dt \\
   &= mc^2 \int_{t_1}^{t_2} \left[\beta^2 \frac{\beta \frac{d\beta}{dt}}{(1-\beta^2)^{3/2}} + \frac{\beta \frac{d\beta}{dt}(1-\beta^2)}{(1-\beta^2)^{3/2}} \right] dt
   \end{aligned}$$

$$= mc^2 \int_{t_1}^{t_2} \left[\frac{\beta \frac{d\beta}{dt}}{(1-\beta^2)^{3/2}} \right] dt$$

$$= mc^2 \int_{u_1}^{u_2} \frac{\beta \, d\beta}{(1-\beta^2)^{3/2}}$$

$$= mc^2 \int_{u_1}^{u_2} d\left[\frac{1}{(1-\beta^2)^{1/2}} \right]$$

$$= mc^2 \left[\frac{1}{(1-\beta^2)^{1/2}} \right]_{u_1}^{u_2}$$

$$= \frac{mc^2}{\sqrt{1-(u_2/c)^2}} - \frac{mc^2}{\sqrt{1-(u_1/c)^2}}$$

where the limits of the integral in the 10th-12th lines are justified by the fact that β is a function of u. We should also note that we used some of the identities that we derived in § 1.4.

2. Discuss the generalization (or extension) of the concept of mass in Lorentz mechanics.
   Answer: If we start from classical mechanics, then mass means the material content of the massive object and hence it does not include any form of energy since in classical mechanics matter and energy are totally different physical entities. When we come to Lorentz mechanics, we face two stages of generalization (or extension) to the concept of mass:
   • The first is based on extending the concept of rest mass to include all forms of non-kinetic energy where this extension is based on the Poincare mass-energy equivalence relation. This extension can be seen in the definition of total energy as the sum of rest energy (which is equivalent to the rest mass) and kinetic energy.
   • The second is based on extending the concept of mass to include even the kinetic energy, and hence we have rest mass (that excludes kinetic energy) and non-rest mass (that includes kinetic energy). This extension can be seen in the old convention of Lorentz mechanics in defining the non-rest mass (which we may call Lorentzian mass) as: $m = \gamma m_0$ where m_0 is the rest mass. It may also be seen (as a possibility) in the definition of total energy: $E_t = \gamma m_0 c^2$ which is based on the concept of non-rest mass since the difference between the two is a multiplicative constant (i.e. c^2).

5.3 Physical Relations

5.3.1 Newton's Second Law

1. Starting from the generic expression of Newton's second law in its vector form, develop and derive the Lorentzian version of Newton's second law in its 1D form.
   Answer: The generic expression of Newton's second law in its vector form is given by $\mathbf{f} = d\mathbf{p}/dt$. For a 1D motion along the x direction, the velocity of the observed object

5.3.2 Mass-Energy Relation

is $\mathbf{u} = \mathbf{u}(t) = (u_x, 0, 0)$. Now, since the motion is only one dimensional along the x direction, then the y and z components of the force should be zero. Accordingly, the above vector form of Newton's second law will be given in components form by:

$$f_x = \frac{dp_x}{dt} = \frac{d}{dt}(\gamma m u_x) = \gamma^3 m a_x \tag{5}$$

$$f_y = 0 \tag{6}$$

$$f_z = 0 \tag{7}$$

where the Lorentzian definition of p_x is used. Regarding the above expression for the x component of the force (i.e. $f_x = \gamma^3 m a_x$), it can be easily derived as follows:

$$\begin{aligned}
f_x &= \frac{d}{dt}(\gamma m u_x) \tag{8} \\
&= m u_x \frac{d\gamma}{dt} + m\gamma \frac{du_x}{dt} \\
&= m u_x \gamma^3 \beta \frac{d\beta}{dt} + m\gamma a_x \\
&= m u_x \gamma^3 \frac{u_x}{c} \frac{d}{dt}\left(\frac{u_x}{c}\right) + m\gamma a_x \\
&= m\gamma^3 \frac{u_x^2}{c^2} \frac{du_x}{dt} + m\gamma a_x \\
&= m\gamma^3 \beta^2 a_x + m\gamma a_x \\
&= m\gamma \left(\gamma^2 \beta^2\right) a_x + m\gamma a_x \\
&= m\gamma \left(\gamma^2 - 1\right) a_x + m\gamma a_x \\
&= m\gamma^3 a_x - m\gamma a_x + m\gamma a_x \\
&= \gamma^3 m a_x
\end{aligned}$$

where the product rule of differentiation is used in the second line. We also used some identities that have been derived in § 1.4. As seen, u_x and γ (which is a function of u_x) are treated as variable while m is treated as constant.

5.3.2 Mass-Energy Relation

1. Using a Lorentzian argument, justify the factors $(1 + u/c)$ and $(1 - u/c)$, which appear in the first method for deriving the mass-energy relation, as the blue and red shift factors.
   Answer: For an approaching observer-source, the Lorentzian Doppler frequency shift (see § 4.2.8) is given by:

$$\nu = \nu_0 \sqrt{\frac{c+u}{c-u}} = \nu_0 \frac{c+u}{c\sqrt{1-(u/c)^2}} \simeq \nu_0 \frac{c+u}{c} = \nu_0 \left(1 + \frac{u}{c}\right)$$

   where the second step is obtained by multiplying the numerator and denominator with $\sqrt{c+u}$ and the approximation in the third step is based on the assumption that $u \ll c$.

Similarly, for a receding observer-source, the Lorentzian Doppler frequency shift (see § 4.2.8) is given by:

$$\nu = \nu_0 \sqrt{\frac{c-u}{c+u}} = \nu_0 \frac{c-u}{c\sqrt{1-(u/c)^2}} \simeq \nu_0 \frac{c-u}{c} = \nu_0 \left(1 - \frac{u}{c}\right)$$

where the second step is obtained by multiplying the numerator and denominator with $\sqrt{c-u}$ and the approximation in the third step is based on the assumption that $u \ll c$. We note that the energy is proportional to the frequency according to Planck's relation: $E = h\nu$ where h is Planck's constant.

2. Using a classical argument, justify the factors $(1 \pm u/c)$, which appear in the first method for deriving the mass-energy relation, as the blue and red shift factors.

   Answer: As explained in the exercises of § 4.2.8, the equation of the classical Doppler frequency shift is given by:

   $$\nu = \frac{c + v_r}{c + v_s} \nu_0$$

   If we now assume that $v_s = 0$ and $v_r = \pm u$ (where the signs correspond to the approaching and receding receiver which is represented by the observer O in the mass-energy argument) then we have:

   $$\nu = \frac{c \pm u}{c} \nu_0 = \nu_0 \left(1 \pm \frac{u}{c}\right)$$

   As we see, this is based on a stationary source and moving receiver relative to a presumed classical medium of propagation, and this is obviously arbitrary and inconsistent with the presumed physical situation of the problem. We note, as in the previous exercise, that the energy is proportional to the frequency.

3. Show that $\gamma \left(1 \pm \frac{u}{c}\right)$, which appear in the second method for deriving the mass-energy relation, are the Lorentzian blue and red shift factors.

   Answer: For an approaching observer-source, the Lorentzian Doppler frequency shift (see § 4.2.8) is given by:

   $$\nu = \nu_0 \sqrt{\frac{c+u}{c-u}} = \nu_0 \frac{c+u}{c\sqrt{1-(u/c)^2}} = \nu_0 \gamma \frac{c+u}{c} = \nu_0 \gamma \left(1 + \frac{u}{c}\right)$$

   where the second step is obtained by multiplying the numerator and denominator with $\sqrt{c+u}$.

   Similarly, for a receding observer-source, the Lorentzian Doppler frequency shift (see § 4.2.8) is given by:

   $$\nu = \nu_0 \sqrt{\frac{c-u}{c+u}} = \nu_0 \frac{c-u}{c\sqrt{1-(u/c)^2}} = \nu_0 \gamma \frac{c-u}{c} = \nu_0 \gamma \left(1 - \frac{u}{c}\right)$$

   where the second step is obtained by multiplying the numerator and denominator with $\sqrt{c-u}$. As indicated already, the energy is proportional to the frequency.

5.3.3 Momentum-Energy Relation

1. Derive the momentum-energy relation of Lorentz mechanics by combining the momentum and total energy formulae.
   Answer: We have:
   $$E = mc^2\gamma \quad \Rightarrow \quad \frac{E}{mc^2} = \gamma$$
   $$p = mu\gamma \quad \Rightarrow \quad \frac{p}{mc} = \gamma\beta$$
   Hence:
   $$\left(\frac{E}{mc^2}\right)^2 - \left(\frac{p}{mc}\right)^2 = \gamma^2 - \gamma^2\beta^2$$
   $$\left(\frac{E}{mc^2}\right)^2 - \left(\frac{p}{mc}\right)^2 = 1$$
   $$\frac{E^2}{m^2c^4} - \frac{p^2}{m^2c^2} = 1$$
   $$E^2 - p^2c^2 = m^2c^4$$
   $$E^2 = p^2c^2 + m^2c^4$$
   where the second line is based on one of the identities of § 1.4.

2. What you observe from a close inspection of the above three methods (two given in the text and one in the previous exercise) for deriving the momentum-energy relation?
   Answer: They use a combination of the total energy formula and the momentum formula of Lorentz mechanics where this combination is done in slightly different ways and at different stages. This should come as no surprise since the derived formula is the "momentum-energy" relation and hence the momentum relation and the energy relation are usually needed in these derivations. So, these methods of derivation are the same in essence although they follow different ways of mathematical casting and manipulation. Regarding the time dilation triangle method, it may be regarded as an independent method of derivation although it also uses the Lorentzian definition of momentum and energy to obtain the final form of the momentum-energy relation. This may also be claimed to be the case with the method of § 6.5.5 but this should be less justified due to its prime reliance on the Lorentzian definition of momentum and energy.

5.3.4 Work-Energy Relation

5.4 Conservation Laws

1. Prove the relation: $p = m\frac{dx}{d\tau}$ which is given in the solved problems.
   Answer: This relation can be proved as follows:
   $$\gamma_u = \frac{dt}{d\tau}$$

5.4 Conservation Laws

$$\gamma_u \frac{dx}{dt} = \frac{dx}{dt}\frac{dt}{d\tau}$$

$$\gamma_u \frac{dx}{dt} = \frac{dx}{d\tau}$$

$$\gamma_u u = \frac{dx}{d\tau}$$

$$\gamma_u m u = m\frac{dx}{d\tau}$$

$$p = m\frac{dx}{d\tau}$$

where in the first line we used the time dilation formula (see § 6.2) while in the third line we used the chain rule of differentiation.

2. Prove the relation: $\frac{dt'}{d\tau} = \frac{E'}{mc^2}$ which is given in the solved problems.
   Answer: This relation is no more than a combination of the time dilation formula (i.e. $\frac{dt'}{d\tau} = \gamma$) plus the total energy formula (i.e. $E' = mc^2\gamma$ and hence $\frac{E'}{mc^2} = \gamma$), that is:

$$\frac{dt'}{d\tau} = \gamma = \frac{E'}{mc^2}$$

3. Show that the conservation of kinetic energy is Lorentz invariant across all inertial frames, i.e. if the kinetic energy is conserved in one inertial frame then it is conserved in all inertial frames (and if not it is not).[97]
   Answer: It was shown earlier that the conservation of total energy is Lorentz invariant across all inertial frames. Now, since the rest energy (or the rest mass) is Lorentz invariant and the total energy is the sum of the rest energy and the kinetic energy then the conservation of kinetic energy should also be Lorentz invariant. More formally, we have:

$$E = E_0 + E_k \qquad \rightarrow \qquad E_k = E - E_0$$

Now, since the conservation of E is Lorentz invariant and E_0 is Lorentz invariant then the conservation of E_k should be Lorentz invariant. We should remind the reader that if E really represents the total energy then all forms of non-kinetic energy should be included in the rest energy due to the equivalence between mass and energy (refer to § 4.1.8, § 4.3.2 and § 5.3.2).

4. Discuss the issue that have been indicated in several places in the book, and in this chapter in particular, that is many of the arguments and proofs that are used to establish the theory of Lorentz mechanics are not sufficiently rigorous.[98]
   Answer: The obvious implication of this is that even though the Lorentzian results

---

[97] This question is not about the conservation of kinetic energy, which is untrue in general, but about the invariance of this conservation, i.e. assuming it is conserved in a given inertial frame (as indicated by "if not it is not"). It should also be obvious that this question is not about the invariance of kinetic energy which is obviously untrue.

[98] In fact, this lack of rigor applies even to some arguments that we presented in this section. However, our purpose of these arguments is to outline the theoretical foundations of these principles rather than proving them rigorously, and hence they are more sensible and useful as pedagogical demonstrations

5.4 Conservation Laws

that are obtained from these arguments and proofs are experimentally supported, they cannot be regarded as theoretically established facts because their theoretical basis within the framework of Lorentz mechanics is not well established. This means that these results may be regarded as empirical with no sufficient theoretical support. This should allow the possibility of a more perfect and general theory that may replace the theory of Lorentz mechanics where the theory of Lorentz mechanics could become an approximation to that theory and hence the experimental evidence actually belongs to the other theory which can provide more rigorous and sound theoretical foundations and arguments for these (as yet) empirical laws.

5. Suggest a different plan for establishing the conservation principles of total energy and momentum and their invariance across all inertial frames.
   Answer: The reader may consider starting from Newton's second law in a particular inertial frame where momentum is conserved in the absence of external forces and extending this to all inertial frames and to total energy. A suitable set of axioms and assumptions should be prepared in advance for this adventure although some of which may emerge (or may be modified) during the theoretical development.

6. Why we need to establish the conservation principles of total energy and momentum plus their invariance across all inertial frames (i.e. why it is not sufficient to establish these conservation principles without their invariance)?
   Answer: It should be obvious that the conservation of total energy and momentum in a particular inertial frame (and even in a number of inertial frames) is not sufficient for establishing these principles in general, and hence the invariance of these conservation principles is still required to establish their general validity. This should have direct practical significance to any scientific work since the observer has access to a limited number of frames and hence he needs this generalization to reach frames in which he cannot verify these principles directly and experimentally.

---

than mathematical and theoretical substantiations. This fact applies to large parts of the theoretical structure of Lorentz mechanics and its alleged proofs, whether those presented in this book or in the wider literature of this subject, especially those related to the conservation laws which are distinguished by their problematic nature.

Chapter 6
Tensor Formulation of Lorentz Mechanics

6.1 Preliminaries

1. Highlight the role of tensor calculus in making the physical laws form invariant.
   Answer: Based on our previous and upcoming investigations, we note that in many cases "being invariant" means that the law can be formulated in an invariant form. Hence, not all the formulations of Lorentz mechanics (as given previously) are actually form invariant since they are formulated in terms of spacetime coordinates of a particular frame although they can be cast in an invariant form. This should highlight the need for tensor calculus to put the physical laws in a mathematically form invariant shape. So in brief, we can say: we have physical form invariance which means that the law is capable of being invariant regardless of its actual form, and mathematical form invariance which means that the actual formulation of the law is form invariant. So, the function of mathematical tools like tensor calculus is to convert the ability of the law to be form invariant to the actuality of being form invariant; i.e. converting the physical invariance to a mathematical invariance. This should explain why some of the tensor formulations are different in form from the physical formulation (and hence we may talk about the tensorial and non-tensorial form of a particular law) because the physical formulations usually reflect the form of the law from the viewpoint of a particular frame and hence it is not necessarily form invariant mathematically.
2. Briefly explain the summation convention that is commonly used in tensor calculus. What about the variance type (i.e. being covariant or contravariant) of the repeated index?
   Answer: The summation convention means that a twice-repeated index in a tensor term implies summation over the range (which represents the dimension of the underlying manifold) of the repeated index. For example, if j ranges over $1, 2, 3$ and μ ranges over $0, 1, 2, 3$ then we have:

$$dX_j dX^j = \sum_{j=1}^{3} dX_j dX^j = dX_1 dX^1 + dX_2 dX^2 + dX_3 dX^3$$

$$dx^\mu dx_\mu = \sum_{\mu=0}^{3} dx^\mu dx_\mu = dx^0 dx_0 + dx^1 dx_1 + dx^2 dx_2 + dx^3 dx_3$$

The repeated indices should in general vary in their variance type and hence one should be covariant and the other contravariant. However, if the manifold is coordinated by an orthonormal Cartesian system[99] then the variance type of the summation indices

---
[99] The metric tensor will then be diagonal with all the diagonal elements being $+1$.

will be irrelevant and hence both can be covariant or contravariant, as well as can be mixed.

3. Show formally that τ really represents the proper time.
   Answer: According to the given definition of τ in its infinitesimal differential form, we have:
   $$d\tau = \frac{\sqrt{|d\sigma^2|}}{c} = \frac{\sqrt{|(dx^0)^2 - (dx^1)^2 - (dx^2)^2 - (dx^3)^2|}}{c} \tag{9}$$
   Now, for a given observer who is in the rest frame of the observed object (call it proper observer) we have $dx^1 = dx^2 = dx^3 = 0$ and hence we have:
   $$d\tau = \frac{\sqrt{|d\sigma^2|}}{c} = \frac{\sqrt{|(dx^0)^2|}}{c} = \frac{cdt}{c} = dt \tag{10}$$
   where t is the proper time of the object (and the given observer). So for this case, τ obviously represents the proper time. Now, since σ is invariant then τ is also invariant and hence it is the proper time of the observed object in all frames.

6.2 Useful Mathematics

1. Briefly define the Laplacian 4-operator.
   Answer: The Laplacian operator in the 4D spacetime manifold, which is also called the d'Alembert operator or the d'Alembertian, is the Laplacian of the Minkowski spacetime which is the space of Lorentz mechanics. This operator is an extension to the ordinary Laplacian, which is a spatial 3-operator, by including the temporal coordinate and hence it is given by:
   $$\Box^2 = \frac{1}{c^2}\frac{\partial^2}{\partial t^2} - \nabla^2 \quad \text{or} \quad \Box^2 = -\frac{1}{c^2}\frac{\partial^2}{\partial t^2} + \nabla^2 \tag{11}$$
   where ∇^2 is the ordinary 3D Laplacian.

2. Using tensor notation and the generic definition of Laplacian in nD space (i.e. divergence of gradient), derive the mathematical expression for the d'Alembertian operator assuming an underlying rectangular Cartesian coordinate system for the spatial part.
   Answer: If h is a scalar field and $g^{\mu\nu}$ is the contravariant metric tensor of the Minkowski spacetime (see § 6.3), then we have:
   $$\begin{aligned}\Box^2 h &\equiv \text{div}\,(\text{grad}\,h) \\ &= \frac{\partial}{\partial x^\mu}\left(g^{\mu\nu}\frac{\partial h}{\partial x^\nu}\right) \\ &= g^{\mu\nu}\frac{\partial^2 h}{\partial x^\mu \partial x^\nu} \\ &= g^{00}\frac{\partial^2 h}{\partial x^0 \partial x^0} + g^{11}\frac{\partial^2 h}{\partial x^1 \partial x^1} + g^{22}\frac{\partial^2 h}{\partial x^2 \partial x^2} + g^{33}\frac{\partial^2 h}{\partial x^3 \partial x^3}\end{aligned}$$

$$= \frac{\partial^2 h}{(\partial x^0)^2} - \frac{\partial^2 h}{(\partial x^1)^2} - \frac{\partial^2 h}{(\partial x^2)^2} - \frac{\partial^2 h}{(\partial x^3)^2}$$

$$= \frac{1}{c^2}\frac{\partial^2 h}{\partial t^2} - \nabla^2 h$$

where the metric tensor in the form $[g^{\mu\nu}] = \text{diag}[1, -1, -1, -1]$ is used in the fourth and fifth steps. We note that treating $g^{\mu\nu}$ as constant in the third step is justified by the fact that the metric tensor for this case is constant (or similarly, the metric tensor is constant with respect to covariant derivative which is equivalent to partial derivative for this case of having a rectilinear system).

3. Using tensor notation and assuming a rectangular Cartesian coordinate system for the spatial part, show that the d'Alembertian operator is invariant under the Lorentz transformations.
   Answer: The d'Alembertian operator can be expressed in tensor notation as:

   $$\Box^2 = g^{\mu\nu}\partial_\nu\partial_\mu$$

   where $\mu, \nu = 0, 1, 2, 3$. Now, since $g^{\mu\nu}$ are constants when the spatial part is coordinated by a rectangular Cartesian system, then $\Box^2$ is invariant.

4. Show that:

   $$\frac{dx^i}{d\tau} = \gamma u^i \qquad (i = 1, 2, 3)$$

   Answer: Using the chain rule of differentiation with the result that we derived in the solved problems, we obtain:

   $$\frac{dx^i}{d\tau} = \frac{dx^i}{dt}\frac{dt}{d\tau} = u^i\gamma = \gamma u^i$$

6.3 Minkowski Metric Tensor

1. Why the Minkowski spacetime is the appropriate "space" for Lorentz mechanics?
   Answer: Because its metric is invariant under the Lorentz spacetime coordinate transformations and hence it ensures the form-invariance property for the physical formulations of Lorentz mechanics.

2. What is the relation between the spacetime interval and the line element of the Minkowski spacetime?
   Answer: They are essentially the same and have the same symbol $d\sigma$. However, the spacetime interval may be used by some to label the quadratic form of the line element, i.e. $(d\sigma)^2$.

3. What "homogeneous coordinate system" means? How can we homogenize the coordinates of the Minkowski spacetime?
   Answer: When all the diagonal elements of a diagonal metric tensor of a flat space are $+1$, the coordinate system is described as homogeneous. The coordinate system of the Minkowski spacetime (with a rectangular Cartesian spatial part) can be homogenized

(i.e. made homogeneous) by defining the temporal coordinate (whether x^0 or x^4) as ict where i is the imaginary unit because all the diagonal elements of the metric tensor will then have the same sign which is plus and hence the quadratic form of the line element will take the simple form:

$$(d\sigma)^2 = dx^\mu dx^\mu$$

where the summation convention applies. The system may also be homogenized by introducing the imaginary unit i on the spatial coordinates instead of the temporal coordinate.

4. What is the significance of the fact that a free particle in the Minkowski spacetime follows a geodesic trajectory?
   Answer: It means that a free particle will follow a straight line in this 4D flat space. Now, if for the sake of simplicity we use the ordinary notation for the space coordinates and time taking $c = 1$ in our unit system, then the trajectory can be represented by the following equations:

$$\frac{x}{A} = \frac{y}{B} = \frac{z}{C} = t$$

where A, B, C are constants. On taking the time derivative of these equations we obtain:

$$\mathbf{u} = (u_x, u_y, u_z) = (A, B, C)$$

i.e. the velocity is constant which is a statement of Newton's first law that a free particle will continue in its state of rest or uniform motion. This is also consistent with the fact that in a flat space (like the Minkowski spacetime) a curve is a geodesic *iff* it is a straight line.

5. Compare the "space trajectory" of a free particle in the 3D ordinary space with its "spacetime trajectory" in the 4D spacetime manifold.
   Answer: Its space trajectory is a straight line in the ordinary sense, while its spacetime trajectory is a straight line in the sense that it follows a straight line in its spatial path with a constant velocity (or speed), i.e. it is straight both in space and in time. So, following a straight trajectory in the spacetime means uniform motion, or indeed obeying Newton's first law of motion. In both cases, the trajectory is a geodesic in the given space.

6. Find the metric tensor and the quadratic form of the Minkowski spacetime with an underlying spatial cylindrical coordinate system.
   Answer: The covariant form of the metric tensor of the cylindrical coordinate system of a 3D space identified by the coordinates (ρ, ϕ, z) is given by:

$$[g_{ij}] = \text{diag}\left[1, \rho^2, 1\right]$$

Hence, if we add the temporal element as a zeroth component, we obtain:

$$[g_{\mu\nu}] = \text{diag}\left[-1, 1, \rho^2, 1\right]$$

where $\mu, \nu = 0, 1, 2, 3$. Accordingly, the quadratic form of the Minkowski spacetime is given by:

$$(d\sigma)^2 = -(cdt)^2 + (d\rho)^2 + \rho^2 (d\phi)^2 + (dz)^2$$

where the squares of the spacetime coordinate differentials (i.e. $(cdt)^2$, $(d\rho)^2$, $(d\phi)^2$ and $(dz)^2$) are multiplied by their corresponding diagonal elements of the metric tensor (i.e. -1, 1, ρ^2 and 1). The metric tensor may also be given as:

$$[g_{\mu\nu}] = \text{diag}\left[1, -1, -\rho^2, -1\right]$$

and hence the quadratic form will be:

$$(d\sigma)^2 = (cdt)^2 - (d\rho)^2 - \rho^2 (d\phi)^2 - (dz)^2$$

The temporal component may also be regarded as the fourth component and hence the order of the temporal and spatial elements and terms in the above two forms will be reversed (corresponding to $\mu, \nu = 1, 2, 3, 4$).

7. Repeat the previous question with an underlying spatial spherical coordinate system.
   Answer: The covariant form of the metric tensor of the spherical coordinate system of a 3D space identified by the coordinates (r, θ, ϕ) is given by:

$$[g_{ij}] = \text{diag}\left[1, r^2, r^2 \sin^2 \theta\right]$$

By adding the temporal element as a zeroth component we obtain the metric tensor of the 4D spacetime, that is:

$$[g_{\mu\nu}] = \text{diag}\left[-1, 1, r^2, r^2 \sin^2 \theta\right]$$

where $\mu, \nu = 0, 1, 2, 3$. Accordingly, the quadratic form of the Minkowski spacetime is given by:

$$(d\sigma)^2 = -(cdt)^2 + (dr)^2 + r^2 (d\theta)^2 + r^2 \sin^2 \theta (d\phi)^2$$

where the squares of the spacetime coordinate differentials (i.e. $(cdt)^2$, $(dr)^2$, $(d\theta)^2$ and $(d\phi)^2$) are multiplied by their corresponding diagonal elements of the metric tensor (i.e. -1, 1, r^2 and $r^2 \sin^2 \theta$). The signs of the elements of the metric tensor and the terms of the quadratic form may be reversed and the order of the temporal and spatial coordinates may be shifted as in the previous question.

8. Express the invariance of the quadratic form (and hence the invariance of the spacetime interval) using tensor notation.
   Answer: Using the ordinary finite form (i.e. σ^2), the invariance of the quadratic form between unprimed frame and primed frame can be expressed as:

$$g_{\mu\nu} x^\mu x^\nu = g'_{\mu\nu} x'^\mu x'^\nu$$

where the indexed g represent the elements of the spacetime metric tensor while the indexed x represent general spacetime coordinates of the unprimed and primed frames, and $\mu, \nu = 0, 1, 2, 3$ (or $\mu, \nu = 1, 2, 3, 4$).

9. Compare the Minkowski spacetime with an ordinary 4D Euclidean space.
   Answer: Some points are:
   - The quadratic form of an ordinary 4D Euclidean space is positive definite (assuming

non-trivial interval), while the quadratic form of Minkowski spacetime is not, i.e. it can be positive (timelike) or negative (spacelike) or zero (lightlike).
• The metric of an ordinary 4D Euclidean space can be represented by a diagonal metric tensor where all its diagonal elements are positive of unity magnitude (i.e. +1), while the metric of Minkowski spacetime can be represented by a diagonal metric tensor with mixed positive and negative diagonal elements of unity magnitude (i.e. ±1).

6.4 Lorentz Transformations in Matrix and Tensor Form

1. Show that $\mathbf{L}$ and $\mathbf{L}^{-1}$ are inverses of each other by verifying the relations: $\mathbf{L}\mathbf{L}^{-1} = \mathbf{L}^{-1}\mathbf{L} = \mathbf{I}$.
   Answer: For the relation $\mathbf{L}\mathbf{L}^{-1} = \mathbf{I}$ we have:

$$\mathbf{L}\mathbf{L}^{-1} = \begin{bmatrix} \gamma & -\beta\gamma & 0 & 0 \\ -\beta\gamma & \gamma & 0 & 0 \\ 0 & 0 & 1 & 0 \\ 0 & 0 & 0 & 1 \end{bmatrix} \begin{bmatrix} \gamma & \beta\gamma & 0 & 0 \\ \beta\gamma & \gamma & 0 & 0 \\ 0 & 0 & 1 & 0 \\ 0 & 0 & 0 & 1 \end{bmatrix}$$

$$= \begin{bmatrix} \gamma^2 - \gamma^2\beta^2 & \beta\gamma^2 - \beta\gamma^2 & 0 & 0 \\ -\beta\gamma^2 + \beta\gamma^2 & -\gamma^2\beta^2 + \gamma^2 & 0 & 0 \\ 0 & 0 & 1 & 0 \\ 0 & 0 & 0 & 1 \end{bmatrix}$$

$$= \begin{bmatrix} 1 & 0 & 0 & 0 \\ 0 & 1 & 0 & 0 \\ 0 & 0 & 1 & 0 \\ 0 & 0 & 0 & 1 \end{bmatrix}$$

$$= \mathbf{I}$$

   where the identity $\gamma^2 - \gamma^2\beta^2 = 1$ is used (refer to § 1.4).
   The relation $\mathbf{L}^{-1}\mathbf{L} = \mathbf{I}$ can be obtained by a similar procedure. However, it can be obtained directly from the relation $\mathbf{L}\mathbf{L}^{-1} = \mathbf{I}$ by the known rules of linear algebra about matrix inverse.

2. Express the relation $\mathbf{L}\mathbf{L}^{-1} = \mathbf{L}^{-1}\mathbf{L} = \mathbf{I}$ in tensor form.
   Answer: If $\mathbf{L}^{-1} \equiv \mathbf{M}$ then we have:

$$L^\mu_\omega M^\omega_\nu = M^\mu_\omega L^\omega_\nu = \delta^\mu_\nu$$

   where the indexed δ is the Kronecker delta tensor.

6.5 Vector, Tensor and Matrix Formulation

6.5.1 Spacetime Position and Displacement 4-Vector

6.5.2 Quadratic Form of Spacetime Interval

1. Justify the invariance of the quadratic form of the spacetime interval by a simple reason.
   Answer: The quadratic form of the spacetime interval is the inner product of a 4-

vector (i.e. position in ordinary form and displacement in differential form) by itself (i.e. covariant by contravariant) and hence it is Lorentz invariant (see § 6.1).

6.5.3 Velocity

1. What are the components of the contravariant velocity 4-vector?
   Answer: We have:
   $$\mathbf{U} = \frac{d}{d\tau}\left(x^0, x^1, x^2, x^3\right) = \frac{d}{d\tau}(ct, x, y, z) = (c\gamma, \gamma u_x, \gamma u_y, \gamma u_z) = \gamma(c, u_x, u_y, u_z)$$
   where the identities $\frac{dt}{d\tau} = \gamma$ and $\frac{dx^i}{d\tau} = \gamma u^i$ are used (see § 6.2).

2. What is the modulus of the velocity 4-vector? Comment on the result.
   Answer: From the result of the previous question, we have:
   $$\begin{aligned}
   \sqrt{|\mathbf{U}^2|} &= \sqrt{|\mathbf{U} \cdot \mathbf{U}|} \\
   &= \sqrt{|U^\mu U_\mu|} \\
   &= \sqrt{|\gamma(c, u_x, u_y, u_z) \cdot \gamma(c, -u_x, -u_y, -u_z)|} \\
   &= \sqrt{\gamma^2\left(c^2 - u_x^2 - u_y^2 - u_z^2\right)} \\
   &= \sqrt{\gamma^2\left(c^2 - u^2\right)} \\
   &= \gamma c\sqrt{1 - u^2/c^2} \\
   &= \gamma c \frac{1}{\gamma} \\
   &= c
   \end{aligned}$$
   where $u < c$.
   Comment: this is inline with the proposal that a change of velocity in the Lorentzian space is equivalent to a rotation through an angle due to the contraction of spacetime coordinates under the influence of motion. Also refer to the exercises of § 4.2.2. We note that the modulus (being equal to the constant c) is Lorentz invariant as it should be since it is the inner product of the velocity 4-vector by itself.

6.5.4 Acceleration

1. Define the acceleration 4-vector assuming a general curvilinear coordinate system.
   Answer: For general curvilinear coordinate systems, the acceleration 4-vector is defined as the absolute (or intrinsic) derivative of the velocity 4-vector with respect to τ and hence it is given by:
   $$A^\mu = \frac{\delta U^\mu}{\delta \tau} = \frac{d^2 x^\mu}{d\tau^2} + \Gamma^\mu_{\nu\omega} \frac{dx^\nu}{d\tau} \frac{dx^\omega}{d\tau}$$
   where the indexed Γ is the Christoffel symbol of the second kind for the 4D space.

2. Referring to the previous question, what is the equation of geodesics in the Minkowski spacetime?
   Answer: The geodesics in the Minkowski spacetime are characterized by zero acceleration. Hence, the geodesics in the Minkowski spacetime are represented by the equation:
$$\frac{d^2 x^\mu}{d\tau^2} + \Gamma^\mu_{\nu\omega} \frac{dx^\nu}{d\tau} \frac{dx^\omega}{d\tau} = 0$$

3. Assuming that the Minkowski spacetime is coordinated by a rectangular Cartesian system, how the equation of geodesics in the Minkowski spacetime will simplify?
   Answer: For the Minkowski metric (with rectangular Cartesian system) the Christoffel symbols vanish identically, and hence the equation of geodesics in the Minkowski spacetime will be simplified to:
$$\frac{d^2 x^\mu}{d\tau^2} = 0$$

6.5.5 Momentum

1. Show that the rest energy is proportional to the length of the momentum 4-vector. What you conclude?
   Answer: The length of the momentum 4-vector is $\sqrt{|\mathbf{P}^2|}$. Using the results obtained in the solved problems, we have:

$$\begin{aligned}
\mathbf{P}^2 &= \mathbf{P} \cdot \mathbf{P} \\
&= P_\mu P^\mu \\
&= (m_0 \gamma c)^2 - (m_0 \gamma u_x)^2 - (m_0 \gamma u_y)^2 - (m_0 \gamma u_z)^2 \\
&= m_0^2 \gamma^2 \left(c^2 - u_x^2 - u_y^2 - u_z^2 \right) \\
&= m_0^2 \gamma^2 \left(c^2 - u^2 \right) \\
&= m_0^2 \gamma^2 c^2 \left(1 - \frac{u^2}{c^2} \right) \\
&= m_0^2 \gamma^2 c^2 \frac{1}{\gamma^2} \\
&= m_0^2 c^2 \\
&= \frac{E_0^2}{c^2}
\end{aligned}$$

Hence:
$$\sqrt{|\mathbf{P}^2|} = m_0 c = \frac{E_0}{c}$$

Conclusion: since the rest energy is proportional to the length of a 4-vector, then it is Lorentz invariant.[100] This should also apply to the rest mass. In fact, this serves as a

---
[100] We may also put it in a different way by saying: the length (being equal to the invariant quantity $m_0 c$) is Lorentz invariant as it should be since it is the inner product of the momentum 4-vector by itself.

6.5.5 Momentum

consistency check.[101]

2. Using the result of the previous exercise and the definition of the momentum 4-vector, as well as other previously-given standard definitions, derive the momentum-energy relation.
Answer: We have:

$$\begin{aligned}
\mathbf{P}^2 &= P_\mu P^\mu \\
\mathbf{P}^2 &= (m_0\gamma c)^2 - (m_0\gamma u_x)^2 - (m_0\gamma u_y)^2 - (m_0\gamma u_z)^2 \\
\mathbf{P}^2 &= \left(\frac{E}{c}\right)^2 - m_0^2\gamma^2\left(u_x^2 + u_y^2 + u_z^2\right) \\
\mathbf{P}^2 &= \left(\frac{E}{c}\right)^2 - m_0^2\gamma^2 u^2 \\
m_0^2 c^2 &= \left(\frac{E}{c}\right)^2 - p^2 \\
m_0^2 c^4 &= E^2 - p^2 c^2 \\
E^2 &= p^2 c^2 + m_0^2 c^4
\end{aligned}$$

which is the momentum-energy relation (refer to § 4.3.3 and § 5.3.3) noting that in the present subsection we follow the old convention about mass (and hence the rest mass is labeled with m_0) while in the previous subsections we followed the modern convention about mass (and hence the mass in general including rest mass was labeled with m).

3. How is the momentum 4-vector transformed between inertial frames in standard setting?
Answer: Using the transformation equation of 4-vectors (i.e. $A'^\mu = L^\mu_\nu A^\nu$ with L^μ_ν being given in matrix form in § 6.4), we obtain:

$$\begin{bmatrix} P'^0 \\ P'^1 \\ P'^2 \\ P'^3 \end{bmatrix} = \begin{bmatrix} \gamma & -\beta\gamma & 0 & 0 \\ -\beta\gamma & \gamma & 0 & 0 \\ 0 & 0 & 1 & 0 \\ 0 & 0 & 0 & 1 \end{bmatrix} \begin{bmatrix} P^0 \\ P^1 \\ P^2 \\ P^3 \end{bmatrix} = \begin{bmatrix} \gamma P^0 - \beta\gamma P^1 \\ -\beta\gamma P^0 + \gamma P^1 \\ P^2 \\ P^3 \end{bmatrix}$$

This is equivalent to (refer to the solved problems):

$$\begin{bmatrix} E' \\ p'_x \\ p'_y \\ p'_z \end{bmatrix} = \begin{bmatrix} \gamma E - v\gamma p_x \\ -\beta\gamma(E/c) + \gamma p_x \\ p_y \\ p_z \end{bmatrix}$$

We note that this is consistent with what we found previously (see § 5.2.1) that the momentum spatial components in the perpendicular directions to the direction of motion (i.e. y and z according to the standard setting) are Lorentz invariant across inertial frames.

---

[101] This result may also be interpreted (arguably) as another demonstration of the energy-momentum conservation where the rest mass prior to any interaction is a fixed quantity.

6.5.6 Force and Newton's Second Law

1. Using the given tensor formulations, show that the world line of a free massive particle is a straight line in the 4D Minkowski spacetime.
   Answer: We have:
   $$F^\mu = \frac{dP^\mu}{d\tau} = m_0 \frac{d^2 x^\mu}{d\tau^2}$$
   For a massive particle to be free we should have $F^\mu = 0$. Now, since m_0 cannot vanish for massive particles then the condition $F^\mu = 0$ implies:
   $$\frac{d^2 x^\mu}{d\tau^2} = 0$$
   which is the equation of a straight line in a 4D space (refer to the exercises of § 6.3). As discussed earlier, this means that the world line of a free particle is a geodesic in the Minkowski spacetime which is Newton's first law.

2. Give a tensor form of Newton's second law assuming a curvilinear coordinate system.
   Answer: A tensor form of Newton's second law in a frame with a curvilinear coordinate system is obtained by replacing the ordinary total derivative in the above definition of the force 4-vector by the intrinsic derivative and hence it is given by:
   $$F^\mu = \frac{\delta P^\mu}{\delta \tau}$$

3. Show that a uniformly accelerated one dimensional motion is equivalent to a constant Lorentz force.
   Answer: For a constant Lorentz force along the x direction, we have:
   $$m_0 \frac{d^2 x}{d\tau^2} = C$$
   where C is a constant. On integrating this equation twice, we obtain a quadratic equation in time which is characteristic for uniformly accelerated motion. The argument can be reversed to obtain the other part of the equivalence.

4. Find the relation between the spatial components of the tensorial force (i.e. $F^i = \frac{dP^i}{d\tau}$) and the non-tensorial force as defined earlier (i.e. $f^i = \frac{dp^i}{dt}$).
   Answer: We have:[102]
   $$\begin{aligned} F^i &= \gamma \frac{d}{dt}\left(m \frac{dx^i}{dt}\right) \\ &= \gamma \frac{d}{dt}\left(\gamma m_0 \frac{dx^i}{dt}\right) \\ &= m_0 \gamma \frac{d}{dt}\left(\gamma \frac{dx^i}{dt}\right) \end{aligned}$$

---
[102] Some of the identities of § 1.4 are used in the following.

6.5.7 Electromagnetism and Maxwell's Equations

$$
\begin{aligned}
&= m_0\gamma\left[\frac{d\gamma}{dt}\left(\frac{dx^i}{dt}\right)+\gamma\frac{d^2x^i}{dt^2}\right] \\
&= m_0\gamma\left[\gamma^3\frac{u^i}{c^2}\frac{du^i}{dt}u^i+\gamma a^i\right] \\
&= m_0\gamma\left[\gamma^3\frac{u^i}{c^2}a^iu^i+\gamma a^i\right] \\
&= m_0\gamma\left[\gamma^3\frac{(u^i)^2}{c^2}a^i+\gamma a^i\right] \\
&= m_0\gamma^2 a^i\left[\gamma^2\frac{(u^i)^2}{c^2}+1\right] \\
&= m_0\gamma^2 a^i\gamma^2 \\
&= \gamma\left(\gamma^3 m_0 a^i\right) \\
&= \gamma\left(\frac{dp^i}{dt}\right) \\
&= \gamma f^i
\end{aligned}
$$

where F^i and f^i are the tensorial and non-tensorial force. This can also be shown more briefly as:

$$F^i = \frac{dP^i}{d\tau} = \frac{dP^i}{dt}\frac{dt}{d\tau} = \frac{dp^i}{dt}\frac{dt}{d\tau} = f^i\gamma$$

6.5.7 Electromagnetism and Maxwell's Equations

1. Show that:
$$S'^{\psi\omega} = L^{\psi}_{\mu}L^{\omega}_{\nu}S^{\mu\nu}$$

where the tensors involved are given in the text.
Answer: In matrix form we have:

$$\mathbf{S}' = \left[S'^{\psi\omega}\right] = \left[L^{\psi}_{\mu}\right]\left[S^{\mu\nu}\right]\left[L^{\omega}_{\nu}\right] = \mathbf{LSL}$$

$$
\begin{aligned}
\mathbf{LS} &= \begin{bmatrix} \gamma & -\beta\gamma & 0 & 0 \\ -\beta\gamma & \gamma & 0 & 0 \\ 0 & 0 & 1 & 0 \\ 0 & 0 & 0 & 1 \end{bmatrix} \begin{bmatrix} 0 & E_x/c & E_y/c & E_z/c \\ -E_x/c & 0 & B_z & -B_y \\ -E_y/c & -B_z & 0 & B_x \\ -E_z/c & B_y & -B_x & 0 \end{bmatrix} \\
&= \begin{bmatrix} \beta\gamma E_x/c & \gamma E_x/c & \gamma E_y/c - \beta\gamma B_z & \gamma E_z/c + \beta\gamma B_y \\ -\gamma E_x/c & -\beta\gamma E_x/c & -\beta\gamma E_y/c + \gamma B_z & -\beta\gamma E_z/c - \gamma B_y \\ -E_y/c & -B_z & 0 & B_x \\ -E_z/c & B_y & -B_x & 0 \end{bmatrix}
\end{aligned}
$$

6.5.7 Electromagnetism and Maxwell's Equations

Now, since matrix multiplication is associative then we have:

$$
\begin{aligned}
\mathbf{LSL} &= (\mathbf{LS})\,\mathbf{L} \\
&= \begin{bmatrix} \frac{\beta\gamma E_x}{c} & \frac{\gamma E_x}{c} & \frac{\gamma E_y}{c} - \beta\gamma B_z & \frac{\gamma E_z}{c} + \beta\gamma B_y \\ -\frac{\gamma E_x}{c} & -\frac{\beta\gamma E_x}{c} & -\frac{\beta\gamma E_y}{c} + \gamma B_z & -\frac{\beta\gamma E_z}{c} - \gamma B_y \\ -\frac{E_y}{c} & -B_z & 0 & B_x \\ -\frac{E_z}{c} & B_y & -B_x & 0 \end{bmatrix} \begin{bmatrix} \gamma & -\beta\gamma & 0 & 0 \\ -\beta\gamma & \gamma & 0 & 0 \\ 0 & 0 & 1 & 0 \\ 0 & 0 & 0 & 1 \end{bmatrix} \\
&= \begin{bmatrix} 0 & \frac{E_x}{c} & \gamma\left(\frac{E_y}{c} - \beta B_z\right) & \gamma\left(\frac{E_z}{c} + \beta B_y\right) \\ -\frac{E_x}{c} & 0 & \gamma\left(B_z - \frac{\beta E_y}{c}\right) & -\gamma\left(B_y + \frac{\beta E_z}{c}\right) \\ -\gamma\left(\frac{E_y}{c} - \beta B_z\right) & -\gamma\left(B_z - \frac{\beta E_y}{c}\right) & 0 & B_x \\ -\gamma\left(\frac{E_z}{c} + \beta B_y\right) & \gamma\left(B_y + \frac{\beta E_z}{c}\right) & -B_x & 0 \end{bmatrix} \\
&= \mathbf{S'}
\end{aligned}
$$

where we used the mathematical identity: $\gamma^2 - \gamma^2\beta^2 = 1$ (see § 1.4).

Chapter 7
Consequences and Predictions of Lorentz Mechanics

7.1 Merging of Space and Time into Spacetime

1. List some examples from Lorentz mechanics that demonstrate the merge of space and time into spacetime.
   Answer: Examples are:
   • Lorentz spacetime coordinate transformations where space and time coordinates in one frame are expressed in terms of both space and time coordinates in the other frame.
   • Invariance of spacetime interval under the Lorenz transformations but not space interval or time interval independently.
   • Tensor formulation of Lorentz mechanics where 3-objects are replaced by 4-objects to reflect the underlying manifold and its invariance properties.

7.2 Length Contraction

1. Summarize the main features of length contraction.
   Answer: The main features of length contraction are outlined in the following points:
   • Length contraction means that the length of an object as measured by an observer who is in a state of relative motion with respect to the object is shorter than its length as measured by an observer who is in the rest frame of the object. Accordingly, the length of an object is longest in its rest frame. Equivalently, the length of physical objects is contracted by motion through spacetime.
   • Length contraction is quantitatively given by $L_0 = \gamma L$ where L_0 and L are the proper and improper length and γ is the Lorentz factor.
   • Length contraction occurs only in the direction of relative motion and hence the dimensions of the object in the perpendicular directions to the direction of motion will keep their proper length.
   • Length contraction can be derived from the Lorentz transformations of spatial coordinates where the measurements of coordinates in the moving frame are assumed to be simultaneous. It may also be postulated to derive these transformations (see § 12.4.2).
   • Length contraction is the 3D spatial part of the 4D contraction of spacetime coordinates by motion.

7.3 Time Dilation

1. How is time dilation effect commonly stated in informal terms?

Answer: It is commonly stated as: moving clocks run slow.

2. What is the essence of length contraction and time dilation?
   Answer: The essence of both these effects is the same that is the coordinates of spacetime contract by the motion where the contraction of the spatial coordinates is represented by length contraction while the contraction of the temporal coordinate is represented by time dilation.

3. Outline some features of time dilation effect.
   Answer: Some features are:
   • Time dilation means that the proper time is shorter than the improper time.[103]
   • Time dilation is quantitatively given by $\Delta t = \gamma \Delta t_0$ where Δt_0 and Δt are the proper and improper time interval and γ is the Lorentz factor.
   • Time dilation can be derived from the Lorentz transformation of the temporal coordinate where the events (i.e. measurements of temporal coordinates) in the transformed frame are assumed to be co-positional. It may also be postulated for the derivation of the Lorentz transformations (see § 12.4.2).
   • Time dilation is the 1D temporal part of the 4D contraction of spacetime coordinates by motion.

7.4 Relativity of Simultaneity

1. Find the condition for two events which are not simultaneous in frame O' to be simultaneous in frame O where O and O' are in a state of standard setting. Repeat the question assuming this time that the two events are simultaneous in frame O' but not in frame O.
   Answer: Let have two events, V_1 and V_2, whose times in frame O' are t'_1 and t'_2 where $t'_1 \neq t'_2$. The times of these events in frame O are t_1 and t_2 where $t_1 = t_2$, that is:

$$t_1 = t_2$$
$$\gamma \left(t'_1 + \frac{vx'_1}{c^2} \right) = \gamma \left(t'_2 + \frac{vx'_2}{c^2} \right)$$
$$t'_1 + \frac{vx'_1}{c^2} = t'_2 + \frac{vx'_2}{c^2}$$
$$\frac{vx'_2}{c^2} - \frac{vx'_1}{c^2} = t'_1 - t'_2$$

where in the second step we used the Lorentz time transformation from frame O' to frame O. So, the condition is:

$$x'_2 - x'_1 = \frac{c^2}{v} (t'_1 - t'_2)$$

---

[103] For our description of time dilation to be more consistent with our description of length contraction, we should say: "the measurement of time interval by a clock is longest when the clock is at rest with respect to the observer". This is based on a different proper-improper perspective from the commonly held perspective on which time dilation is based. In fact, we can call our perspective the time contraction perspective where all spacetime coordinates, whether spatial or temporal, are seen to contract by motion (see § 1.6, § 4.1.2, § 5.1.5 and § 11).

Regarding the second part of the question, if V_1 and V_2 are simultaneous in frame O' then we should have:

$$t'_1 = t'_2$$
$$\gamma\left(t_1 - \frac{vx_1}{c^2}\right) = \gamma\left(t_2 - \frac{vx_2}{c^2}\right)$$
$$t_1 - \frac{vx_1}{c^2} = t_2 - \frac{vx_2}{c^2}$$
$$\frac{vx_2}{c^2} - \frac{vx_1}{c^2} = t_2 - t_1$$

where in the second step we used the Lorentz time transformation from frame O to frame O'. So, the condition in this case is:

$$x_2 - x_1 = \frac{c^2}{v}(t_2 - t_1)$$

In fact, this condition can be obtained more easily from the previous condition by replacing the primed symbols with unprimed symbols and reversing the sign of v.

2. Make a clear distinction between the simultaneity of occurrence and the simultaneity of observation in the context of relativity of simultaneity.
Answer: The simultaneity of occurrence of two events is a global property for a particular frame since any observer in any position in that frame will agree on the simultaneity in this sense. In contrast, the simultaneity of observation is a local property for a particular frame since the simultaneity in this sense depends on the position of the observer in that frame. Hence, we should distinguish between the relativity of simultaneity of occurrence and the fact that events taking place at different locations in a given frame may look to an observer at a particular position in that frame to be in a time order different to their real time order in that frame and hence they can be observed simultaneously by one observer in that frame and non-simultaneously by another observer in that frame or the order between the observations be different. For example, if an observer located at point A in the 2D spacetime diagram of Figure 10 saw an event V_B that occurred at point B where B is one light year away from point A then he might think that this event is occurring "now" and hence it is simultaneous with another event V_A which is really occurring now. He may also think that V_B is occurring one year after another event V_C which occurred one year earlier although V_B and V_C are actually simultaneous in occurrence. So, the difference is that: the simultaneity of occurrence is about the time of occurrence which globally applies throughout any particular frame and hence the relativity of simultaneity of occurrence is about two different frames where time is defined globally in each frame while the simultaneity of observation is about the time of observation of a particular observer located at a particular position in a frame and hence the relativity of simultaneity of observation is about observation of events in a particular frame by different observers that are located in different positions in that frame and hence their observations may disagree in time. To put it in simple terms, the difference is that the relativity of simultaneity of occurrence is about two global

7.4 Relativity of Simultaneity

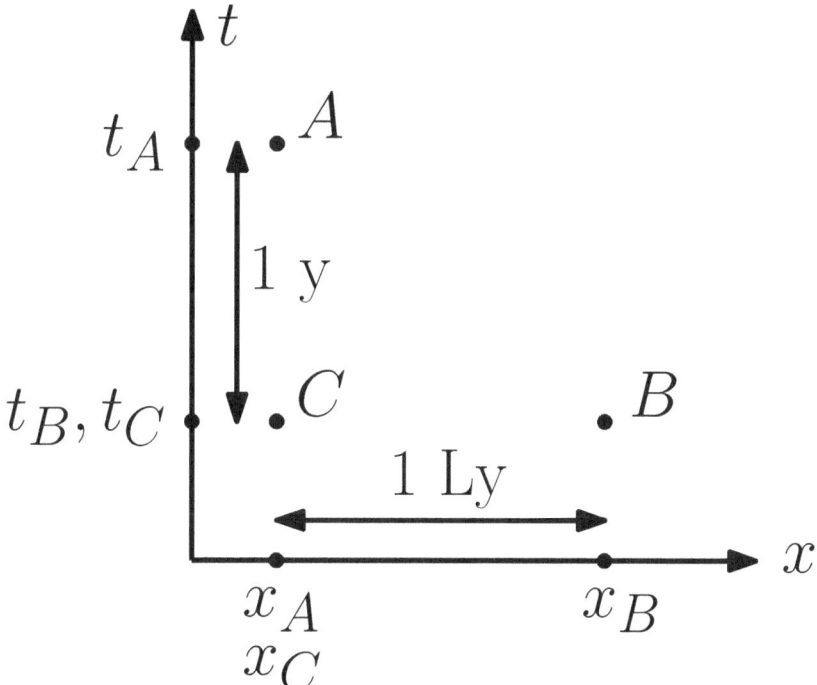

Figure 10: Schematic diagram to illustrate the concept of relativity of simultaneity of observation where we represent events that occur at the same time but in different locations (B and C) and events that occur at the same location but in different times (A and C) as well as events that occur at different locations and times (A and B). We note that Ly stands for a distance of one light year while y stands for a time interval of one year. Also, we used t (instead of ct which represents the temporal coordinate in standard spacetime diagrams) for the purpose of simplicity and to match the explanation given in the text.

observers in two different frames while the relativity of simultaneity of observation is about two local observers in two different locations of the same frame.[104]

3. Show that the relativity of simultaneity of occurrence is proprietary to Lorentz mechanics and hence it does not exist in classical physics.
   Answer: In classical physics the time is transformed as $t = t'$ and hence for two events A and B we have $t_B - t_A = t'_B - t'_A$. So, if the two events are simultaneous in one frame (and hence $t_B - t_A = 0$ or $t'_B - t'_A = 0$) then they must be simultaneous in the other frame as well (i.e. $t'_B - t'_A = 0$ or $t_B - t_A = 0$). Therefore, the relativity of simultaneity of occurrence is proprietary to Lorentz mechanics.
4. O and O' are inertial observers in a state of standard setting. In O frame, events A and B are simultaneous and they are spatially separated by a distance $x_B - x_A = 10^4$ m. Find the time interval between A and B in O' frame if their spatial separation in this frame is $x'_B - x'_A = 10^5$ m.

---

[104] In fact, the simultaneity of observation (and hence the relativity of this simultaneity) may also be envisaged between two frames but this is not needed for our objective of distinguishing between the simultaneity of occurrence and the simultaneity of observation in this context.

Answer: Because A and B are simultaneous in O frame then $t_B - t_A = 0$. From the Lorentz transformation of x coordinate we have:

$$\begin{aligned}
x'_B - x'_A &= \gamma\left[(x_B - x_A) - v(t_B - t_A)\right] \\
x'_B - x'_A &= \gamma\left[(x_B - x_A) - v \times 0\right] \\
10^5 &= 10^4 \gamma \\
\gamma &= 10 \\
v &\simeq 0.9950c
\end{aligned}$$

Now, from the Lorentz transformation of time we have:

$$\begin{aligned}
t'_B - t'_A &= \gamma\left[(t_B - t_A) - \frac{v}{c^2}(x_B - x_A)\right] \\
&= -\gamma\frac{v}{c^2}(x_B - x_A) \\
&\simeq -10 \times \frac{0.9950}{c} \times 10^4 \\
&\simeq -3.3166 \times 10^{-4}\,\text{s}
\end{aligned}$$

where the minus sign indicates the chronological order of the two events in frame O', i.e. B occurred before A. As we see, these events are simultaneous in O frame but they are not simultaneous in O' frame.

5. O and O' are inertial observers in a state of standard setting. In O frame, events A and B are observed to be temporally separated by $t_B - t_A = 0.5$ s and they are spatially separated by a distance $x_B - x_A = 10^{10}$ m while in O' frame they are observed to be simultaneous. What is the relative speed between O and O'?

   Answer: Because they are simultaneous in O' frame then we have $t'_B - t'_A = 0$. From the Lorentz transformation of time we have:

   $$\begin{aligned}
   t'_B - t'_A &= \gamma\left[(t_B - t_A) - \frac{v}{c^2}(x_B - x_A)\right] \\
   0 &= \gamma\left[0.5 - \frac{v}{c^2}10^{10}\right]
   \end{aligned}$$

   Now, since γ cannot be zero then we have:

   $$\begin{aligned}
   0.5 - \frac{v}{c^2}10^{10} &= 0 \\
   v &= \frac{0.5}{10^{10}}c^2 \\
   v &\simeq 4.5 \times 10^6\,\text{m/s}
   \end{aligned}$$

6. Compare simultaneity in classical mechanics and in Lorentz mechanics according to the special relativistic interpretation. Is the denial of absolute time necessary to make sense of the relativity of simultaneity?

   Answer: In classical mechanics, which is based on the existence of an absolute universal time, simultaneity is a simple and obvious concept, that is two events are simultaneous if they occur at the same time. In special relativity there is no absolute universal time since time is a variable associated with an inertial reference frame and hence each particular reference frame has its own time. Therefore, simultaneity is not an absolute and universal concept, that is the two events can be simultaneous for an observer in a particular reference frame but non-simultaneous for another observer in a different reference frame which is in relative uniform motion with respect to the first reference frame. Hence, relativity of simultaneity means that the concept of "simultaneous events taking place at different points in space" has no meaning as such since further qualification is needed to attach a significance to such "simultaneity" by ascribing this to a particular reference frame. As a consequence, two events which are simultaneous for one inertial observer are not necessarily simultaneous to another inertial observer who is in relative motion with respect to the first observer.[105]

The denial of absolute time is not necessary to make sense of the relativity of simultaneity because there could be an absolute time but two frames can disagree on the simultaneity of events when these frames are in a state of relative motion with respect to each other (and hence at least one of these frames is in a state of relative motion with respect to the absolute frame). In brief, the existence of absolute frame (and hence absolute time) is sensible and logically consistent with the concept of relativity of simultaneity where the absolute frame provides an absolute time to all frames although each one of these frames has its own time which may differ from the time of other frames and the time of the absolute frame due to the contraction of spacetime coordinates by the motion relative to the absolute frame. In fact, this is similar to having relativity of simultaneity of observation in a particular frame but we still have "absolute" simultaneity of occurrence in that particular frame where all observers in that frame agree on the simultaneity of occurrence of two events in that frame although they disagree on the simultaneity of observation of these events. So in brief, although different observers in different frames may disagree on the simultaneity of events, they should all agree on the simultaneity or non-simultaneity of events in reference to the absolute time of the absolute frame and hence simultaneity in this sense is absolute.

7.5 Relativity of Co-positionality

1. O and O' are inertial observers in a state of standard setting. In O frame, events A and B are co-positional and they are temporally separated by a time interval $t_B - t_A = 0.01$ s. Assuming that the y and z coordinates of the two events are identical, find the spatial separation between A and B in O' frame if their time separation in this frame is $t'_B - t'_A = 0.02$ s.
   Answer: Because A and B are co-positional in O then $x_B - x_A = 0$. From the Lorentz

---

[105] In brief, if the two events are identical then simultaneity is frame independent and if not then it is frame dependent. Accordingly, "simultaneous" in this context and similar contexts is used in its generic meaning not as opposite to identical.

7.5 Relativity of Co-positionality

transformation of time we have:

$$t'_B - t'_A = \gamma\left[(t_B - t_A) - \frac{v}{c^2}(x_B - x_A)\right]$$
$$t'_B - t'_A = \gamma\left[(t_B - t_A) - \frac{v}{c^2} \times 0\right]$$
$$0.02 = 0.01\gamma$$
$$\gamma = 2$$
$$v = c\sqrt{0.75}$$

Now, from the Lorentz transformation of x coordinate we have:

$$\begin{aligned} x'_B - x'_A &= \gamma\left[(x_B - x_A) - v(t_B - t_A)\right] \\ &= \gamma\left[0 - v(t_B - t_A)\right] \\ &= -\gamma v(t_B - t_A) \\ &= -2 \times c\sqrt{0.75} \times 0.01 \\ &\simeq -5196152\,\text{m} \end{aligned}$$

where the minus sign indicates the spatial order of the two events on the x axis. As we see, these events are co-positional in O frame but they are not co-positional in O' frame (in fact, they are anti-identical).

2. O and O' are inertial observers in a state of standard setting. In O frame, events A and B are observed to be temporally separated by $t_B - t_A = 0.005$ s and they are spatially separated by a distance $x_B - x_A = 10^4$ m while in O' frame they are observed to be co-positional. What is the relative speed between O and O'?
Answer: Because they are co-positional in O' frame then we have $x'_B - x'_A = 0$. From the Lorentz transformation of x coordinate we have:

$$\begin{aligned} x'_B - x'_A &= \gamma\left[(x_B - x_A) - v(t_B - t_A)\right] \\ 0 &= \gamma\left[(x_B - x_A) - v(t_B - t_A)\right] \end{aligned}$$

Now, since γ cannot be zero then we have:

$$\begin{aligned} (x_B - x_A) - v(t_B - t_A) &= 0 \\ v &= \frac{x_B - x_A}{t_B - t_A} \\ v &= \frac{10^4}{0.005} \\ v &= 2 \times 10^6 \,\text{m/s} \end{aligned}$$

3. How do you compare the relativity of simultaneity and the relativity of co-positionality to the absolute simultaneity and absolute co-positionality in reference to the absolute frame? Try to link this to the speed of light as an invariant across all inertial frames.
Answer: We may liken them to the difference between the simultaneity of observation

7.6 Equivalence of Mass and Energy 197

and simultaneity of occurrence (despite the great difference between the two) where the simultaneity of occurrence is the ultimate simultaneity in comparison to the apparent simultaneity of observation similar to the apparent simultaneity of occurrence in a particular frame relative to the absolute simultaneity of occurrence with respect to the absolute frame. In fact, the relativity of simultaneity and the relativity of co-positionality are a cost (against the absolute sense of these concepts) that we pay for adopting the speed of light as the standard for spacetime calibration.

7.6 Equivalence of Mass and Energy

1. Find a common factor between the above consequences and predictions of Lorentz mechanics which are discussed in the sections of this chapter.
   Answer: Apart from the mass-energy equivalence (which may be a classical consequence and not necessarily Lorentzian), all the above consequences (i.e. merging of space and time into spacetime, length contraction, time dilation, relativity of simultaneity and relativity of co-positionality) are related to the merge of space and time into spacetime and the resulting distortion of spatial and temporal coordinates by motion. This should be anticipated since the essence of Lorentz mechanics is the transformation of space and time coordinates in a particular way that is based on the above merge and distortion. In brief, Lorentz mechanics is a theory of space and time and hence all its distinctive consequences should be related to this feature of the theory.
2. Give examples of subsidiary consequences and predictions that can be added to the above list of main consequences and predictions (i.e. merging of space and time into spacetime, length contraction, etc.) of Lorentz mechanics.
   Answer: Examples include the invariance of the speed of light and the invariance of the spacetime interval across inertial frames where these subsidiary consequences and predictions may be seen as results of length contraction and time dilation (or rather spacetime contraction by the effect of motion with the adoption of c for spacetime calibration) which the entire Lorentz mechanics is based upon (see for example § 9.7.1, § 11 and § 12.4.2).

Chapter 8
Evidence for Lorentz Mechanics

8.1 Success of Lorentz Transformations

8.2 Mass-Energy Equivalence

8.3 Prolongation of Lifetime of Elementary Particles

1. An unstable subatomic particle has a proper lifetime $t_0 = 1$ microsecond and a speed $v = 0.99c$. What distance will this particle travel during its lifetime as observed in the laboratory frame?
   Answer: According to the time dilation formula, we have:
   $$t = \gamma t_0 = \frac{1}{\sqrt{1 - 0.99^2}} \simeq 7.089\,\mu s$$
   where t is the improper (or laboratory) lifetime. Hence, the distance d traveled during its lifetime according to the laboratory frame is:
   $$d = vt \simeq 0.99 \times 3 \times 10^8 \times 7.089 \times 10^{-6} \simeq 2127\,\text{m}$$

2. An unstable subatomic particle is observed in the laboratory to travel 300 m at a speed of $v = 0.95c$. What is its proper lifetime? How far the particle will travel according to classical mechanics?
   Answer: We have:
   $$t = \frac{d}{v} = \frac{300}{0.95c}$$
   where t and d are the lifetime and traveled distance according to the laboratory frame. Hence, its proper lifetime t_0 is:
   $$t_0 = \frac{t}{\gamma} = \frac{300}{0.95c}\sqrt{1 - 0.95^2} \simeq 3.287 \times 10^{-7}\,\text{s}$$

   In classical mechanics there is no time dilation and hence the lifetime in the laboratory frame is the same as the lifetime in the particle frame. Accordingly, the distance traveled according to classical mechanics should be:
   $$d_c = vt_0 \simeq 0.95 \times 3 \times 10^8 \times 3.287 \times 10^{-7} \simeq 94\,\text{m}$$
   which is much shorter than the actual (i.e. Lorentzian) distance.

8.3 Prolongation of Lifetime of Elementary Particles

3. A subatomic particle with a proper lifetime $t_0 = 10^{-7}$ s is created in a particle collider and it is moving at a constant velocity $v = 0.7c$ at the instant of its creation. How far the particle will travel according to classical mechanics and according to Lorentz mechanics? Solve the second part once as a time dilation problem and once as a length contraction problem.
Answer: According to classical mechanics, the distance d traveled by the particle is:

$$d = vt_0 \simeq 0.7 \times 3 \times 10^8 \times 10^{-7} = 21\,\mathrm{m}$$

According to Lorentz mechanics, the proper lifetime is the time in the rest frame of the particle, and hence the lifetime in the laboratory frame is prolonged by the Lorentz γ factor according to the time dilation effect. So, if we use the time dilation formula to solve this problem, then the distance d traveled by the particle is:

$$d = vt = v\gamma t_0 \simeq \frac{0.7 \times 3 \times 10^8 \times 10^{-7}}{\sqrt{1 - 0.7^2}} \simeq 29.4\,\mathrm{m}$$

Similarly, if we solve this problem as a length contraction problem, then from the perspective of the rest frame of the particle the distance traveled by the particle is contracted by the Lorentz γ factor due to the length contraction effect and hence the distance in the particle frame is $d_p = d/\gamma$ where d_p and d are the distance in the particle and laboratory frames. So, as seen from the particle frame we have:

$$\begin{aligned} d_p &= vt_0 \\ \frac{d}{\gamma} &= vt_0 \\ d &= v\gamma t_0 \simeq 29.4\,\mathrm{m} \end{aligned}$$

which is identical to what we obtained from using the time dilation formula.

4. An elementary particle with a proper lifetime of 10^{-6} s is observed to travel a distance of 100 m during its lifetime. What is the speed of this particle?
Answer: We have a proper lifetime $t_p = 10^{-6}$ s and a laboratory distance $d_l = 100$ m where the subscripts p and l refer to the particle and laboratory. Hence, we have:

$$\begin{aligned} d_l &= vt_l \\ d_l &= v\gamma t_p \\ v\gamma &= \frac{d_l}{t_p} \\ \beta\gamma &= \frac{d_l}{ct_p} \\ \sqrt{\gamma^2 - 1} &= \frac{d_l}{ct_p} \\ \gamma^2 &= \left(\frac{d_l}{ct_p}\right)^2 + 1 \end{aligned}$$

$$1 - \beta^2 = \left[\left(\frac{d_l}{ct_p}\right)^2 + 1\right]^{-1}$$

$$\beta = \left(1 - \left[\left(\frac{d_l}{ct_p}\right)^2 + 1\right]^{-1}\right)^{1/2}$$

$$v \simeq 0.3162c$$

where in the second step we used the time dilation formula while in the fifth step we used the identity $\beta\gamma = \sqrt{\gamma^2 - 1}$ (see § 1.4).

8.4 Atomic Clock Experiment

1. A clock was placed on board a missile that moves with a constant velocity $v = 1000$ m/s. How long the missile should move (as seen from the frame of a stationary launch pad) for the clock to be 1 millisecond behind an identical clock that stayed stationary on the launch pad?
   Answer: Starting from the time of launch, if we label the time recorded by the missile clock and by the stationary clock with t_m and t_s then from the time dilation formula we have:

$$t_s = \gamma t_m$$
$$t_s - t_m = \gamma t_m - t_m$$
$$t_s - t_m = t_m (\gamma - 1)$$
$$t_m = \frac{t_s - t_m}{\gamma - 1}$$
$$t_m = \frac{0.001}{(1 - 1000^2/c^2)^{-1/2} - 1}$$
$$t_m \simeq 1.8 \times 10^8 \, \text{s}$$
$$t_m \simeq 2083.3 \, \text{days}$$
$$t_m \simeq 5.7 \, \text{years}$$

Now, since the difference between t_s and t_m is very tiny (only 0.001 s in about 5.7 years), we can equate the two times and hence we get:

$$t_s \simeq t_m \simeq 5.7 \, \text{years}$$

which is very long time for this very tiny time difference especially when we consider the velocity which is classically very high. The reported atomic clock experiments do not employ such high speeds or long flight times and hence the presumed time delay is extremely tiny which may cast a shadow on the reliability of this type of experiments especially those conducted several decades ago where the equipment were not as accurate and reliable as the equipment of modern days.

8.5 Stellar Aberration

Chapter 9
Special Relativity

9.1 Characteristic Features of Special Relativity

1. Outline the main features of special relativity.
   Answer: In summary, the theory of special relativity is characterized by the following features:
   • It is based on two postulates which are the postulate of relativity for all inertial frames and the postulate of invariance of the observed speed of light for all inertial observers. The Lorentz spacetime coordinate transformations (and subsequently all the other transformations and formulations of Lorentz mechanics) are derived from these two postulates (see § 12.4.1). Unlike its classical counterpart, the relativity principle in special relativity is based on the denial of the existence of absolute frame of reference, as we will see next. Regarding the invariance of the observed speed of light, the literature of special relativity suggests that the interpretation of this postulate is not the same among the followers of this theory.
   • It is based on denying the existence of absolute space and absolute time and hence denying the existence of any absolute frame of reference. As a consequence, the theory also denies the existence of any medium for the propagation of light such as ether (or at least superfluity of such a medium) since such a medium can provide an absolute frame.
   • The constant c represents a restricted speed to light (which includes other types of electromagnetic radiation and any massless object) and hence no massive object can reach this speed. Moreover, c is the ultimate speed and hence no physical object, whether massive or massless, can exceed this speed. So in brief, the speed of all massive objects must not reach c while the speed of all massless objects must not fall below or exceed c.

2. Can we consider the abolishment of absolute frame (and hence absolute space and absolute time) as an implication of the Lorentz spacetime coordinate transformations and hence as an endorsement to the special relativistic view?
   Answer: No. The Lorentz spacetime coordinate transformations imply the abolishment of absolute frame if we accept the special relativistic interpretation of the relativity principle. In fact, if we postulate this principle in its special relativistic sense then we already assume the abolishment of absolute frame with no need to wait until we derive the Lorentz transformations from this principle according to special relativity. So in brief, the Lorentz spacetime coordinate transformations and all their derived consequences are consistent with the existence of an absolute frame as long as we do not adopt the special relativistic interpretation of the relativity principle.

9.2 Postulates of Special Relativity

1. What are the two postulates of special relativity?
   Answer: They are: the relativity principle whose essence is the equivalence of all inertial frames of reference in their validity for formulating the laws of physics, and the constancy of the observed speed of light in free space for all inertial observers and hence this speed is the same regardless of the state of motion of the source or/and observer.
2. Show by a simple non-rigorous argument that the equivalence of all inertial frames of reference in their validity for formulating the laws of physics implies that it is impossible to detect the state of any inertial frame as being at rest or uniform motion in space in an absolute sense by conducting any experiment in that frame.
   Answer: Since the laws of physics take the same form in all inertial frames of reference, all experiments in any inertial frame will be subject to the same laws and hence all experiments will give similar results as far as the state of rest or uniform motion of the frame is concerned. Hence, it is impossible to detect the state of any inertial frame as being at rest or uniform motion in space in an absolute sense by conducting any experiment in that frame because the laws (and hence the results of these experiments) do not vary depending on the state of rest or uniform motion in space.
   We should remark that if we accept the special relativistic interpretation of the relativity principle, which is based on the denial of the existence of any absolute frame, then this implication can be concluded simply and with no need for any technicality from the fact that the result of any experiment should be independent of the choice of the frame of reference because the laws of physics and their results should not be dependent on our choice and convention about the frames of reference which, in the absence of absolute frame, are purely synthesized mathematical structures and hence they do not represent and they should not affect any physical reality or truth. However, before that we need to find a satisfactory explanation for the fact that acceleration between different frames of reference represents a fundamental physical reality to justify the distinction between inertial and non-inertial frames where the known laws of physics hold true in the former frames but not in the latter frames. As we will see, this distinction puts question marks on the principle of relativity in its special relativistic sense.[106]
3. According to special relativity, the speed of light is independent of the speed of its source and the speed of its observer. Express this premise more efficiently and compactly.
   Answer: Since the relativity principle of special relativity is based on the denial of the existence of absolute frame, where the motion of the light source relative to this frame may have different effect from the motion of the observer relative to this frame, then we can simply say: the essence of the second postulate of special relativity is that the speed of light is independent of the relative motion between the light source and the observer.

---

[106] We should remark that addressing some of these question marks by calling the general theory of relativity does not provide a convincing answer for the abolishment of absolute frame. We hope these issues will be tackled in the future.

9.3 Assessing the Postulates of Special Relativity

9.3.1 Relativity Principle

1. What are the restricted and unrestricted forms of the relativity principle?
   Answer: The essence of the restricted form of the relativity principle is that although there is potentially an absolute frame, this presumed absolute frame does not distinguish between reference frames as long as they are inertial (i.e. they move uniformly relative to this frame) although this frame discriminates against non-inertial frames. This discrimination or distinction takes the form of privileging the inertial frames with holding the laws of physics and denying this privilege from the non-inertial frames. The essence of the unrestricted form of the relativity principle is the denial of the existence of absolute frame and hence the physically-real distinction between inertial and accelerating frames requires explanation. Moreover, the reality of the characteristic Lorentzian effects, like time dilation and length contraction, requires justification.

2. Contemplate on the sufficiency of the restricted form of the relativity principle as a viable explanation for both classical mechanics and Lorentz mechanics.
   Answer: Both classical mechanics and Lorentz mechanics are theories about inertial frames. Now, because the subject of the restricted relativity principle is inertial frames there should be no difference between classical mechanics and Lorentz mechanics in this regard. So, if classical mechanics is compatible with the restricted relativity principle and can be fully explained by the restricted form, then Lorentz mechanics should also be compatible with this principle and can be fully explained by this form with no need to call for a more general form of the relativity principle that abolishes the existence of absolute frame. Accordingly, even if we assume that the distinction between inertial and non-inertial frames and the reality of the characteristic Lorentzian effects can be equally explained by an assumption other than the existence of absolute frame, we still do not need the unrestricted relativity principle to interpret and justify Lorenz mechanics because the restricted relativity principle (with its implication of a potential existence of an absolute frame) is sufficient to accommodate all the new features of Lorentz mechanics and its formalism which are not found in classical mechanics. In brief, the unrestricted relativity principle is at least superfluous if not wrong even if we do not need to assume the existence of absolute frame definitely. In other words, even if we assume that the existence of absolute frame is not needed for the above two reasons, the definite denial of this existence by the unrestricted relativity principle is superfluous and hence it requires justification because any scientific theory should be based on the minimum amount of required assumptions to establish its theoretical structure.

3. Let have two inertial observers, O and O', who are in a state of relative uniform motion. Show that according to the principles of special relativity, each observer will see the clock of the other observer run slow. What you conclude from this?
   Answer: According to the principles of special relativity, there is no absolute frame with respect to which time dilation effect takes place. Hence, the situation of each observer in the eye of the other observer is identical, and therefore time dilation effect should take place in each frame as seen from the other frame due to this symmetrical

9.3.1 Relativity Principle

situation. Accordingly, if we accept the principles of physical reality and truth then time dilation cannot take place in both frames in a real sense because it is contradictory to assume that both the time of O is behind the time of O' and the time of O' is behind the time of O. In other words, as these observers can watch the time of each other during their relative motion, they cannot see any real change in their times, i.e. they remain synchronized. The conclusion is that if we accept special relativity then we should assume that time dilation effect is apparent and not real, i.e. although each observer may notice (or feel wrongly like an illusion) that the clock of the other observer is ticking at a slower rate in an instantaneous sense, they do not feel any real difference between their times over an extended period of their journey. In fact, even if we assume the impossible that is this instantaneous effect accumulates and hence each observer sees the time of the other observer (as recorded by a series of clocks in each frame as described in 3.6) continuously lagging during the journey, this effect cannot be real because of the symmetrical situation which makes it impossible to have t behind t' and t' behind t. So, if we assume that the relative motion between the two frames stopped suddenly and symmetrically and the two observers checked their times, they cannot find that t is lagging behind t' and t' is lagging behind t. In fact, this is related to the twin paradox which will be investigated in § 10.1. The fact that the effect cannot be real will be more obvious when we have three inertial frames which are in relative motion with respect to each other where the time dilation and length contraction will be different in each frame as observed from the other two frames and hence by the principles of reality and truth the effect cannot be real because we cannot have two contradicting real effects. The refute of this challenge by changing frame (or by similar excuses) does not make any sense unless we abolish the principles of physical reality and truth which should lead to the abolishment of science and its purpose since we will then live in a world of illusions and dreams.

4. Investigate all the possibilities for the reality of length contraction and time dilation effects with the relativity postulate of special relativity.
Answer: There are four main possibilities for the reality of length contraction and time dilation effects with the relativity postulate:
• Accepting the relativity postulate of special relativity and considering length contraction and time dilation real effects. This is logically inconsistent because it requires either having asymmetrical effect which is inconsistent with the relativity principle, or having conflicting realities which obviously contradicts the principles of physical reality.
• Rejecting the relativity postulate and considering length contraction and time dilation real effects. This is logically consistent if we assume that length contraction and time dilation effects take place with respect to a preferential reference frame which is the frame of ether or the absolute space.
• Accepting the relativity postulate and considering length contraction and time dilation apparent effects. This apparently means that these effects are errors that require correction.
• Rejecting the relativity postulate and considering length contraction and time dilation apparent effects. Again, this apparently means that these effects are errors that

require correction.
5. What "real" and "apparent" mean in the context of the interpretation of length contraction and time dilation?
   Answer: "Real" means that these effects occur to the observed object itself in the real world, while "apparent" means that these effects are observational phenomena that affect the observation but not the object itself. So, if the contraction of stick is real then its length is shortened, while if it is apparent then the length of the stick did not change but the observed or measured length is shortened (like the immersed stick which is seen broken).

9.3.2 Invariance of Observed Speed of Light

1. Discuss and evaluate the argument that is given in the first solved problem.
   Answer: The point of that argument is that when an observer sees a photon moving with speed c in his frame, then any other observer in another frame should also see the photon moving with speed c in his frame because since the spacetime coordinates are contracted (or dilated) in the second frame compared to the coordinates of the first frame the speed c should also be "contracted" (or "dilated"). In brief, c scales up and down in each frame according to the scaling up and down of the spacetime coordinates in that frame.[107] The essence of this is that c represents an intrinsic scale factor of the spacetime of Lorentz mechanics (because in Lorentz mechanics the calibration of spacetime measurements are based on the speed of light) and hence it scales in proportion with the scaling of spacetime in every frame. The picture will be more clear if we assume the existence of an absolute frame and hence the contraction of spacetime coordinates by the γ factor in any frame is caused by the movement of the frame relative to this absolute frame (see § 1.6). So, if a light signal is observed to have speed c in an inertial frame, then it should also be observed to have speed c in any other inertial frame due to the aforementioned contraction of spacetime (and the scaling of c accordingly) where any velocity component in the signal due to the relative motion will be embedded within the motion of the frame as a whole and hence the scaling of the speed of light by the scaling of spacetime coordinates will not be affected (see § 9.7.1).
2. O and O' are two inertial observers in a state of standard setting. Using the Lorentz spacetime coordinate transformations, show that if O observes the speed of light to be c then O' should also observe the speed of light to be c. Comment on this question.
   Answer: If O observes the speed of light to be c then he should have $x = ct$ where x stands for the distance traveled by the light during the time interval t.[108] On using the Lorentz spacetime coordinate transformations we have:

$$x = ct$$

---

[107] Since the spatial and temporal coordinates are scaled by the same γ factor then the speed of light (which is a ratio of space to time) will be scaled by unity, i.e. it keeps its invariant value c. In other words, if space scales by a factor a and time scales by a factor b then the speed of light will scale by a factor a/b, and because we have $a = b = \gamma$ then the speed will be scaled by 1.

[108] For simplicity and clarity we use x and t instead of Δx and Δt.

9.3.2 Invariance of Observed Speed of Light

$$\gamma\left(x' + vt'\right) = c\gamma\left(t' + \frac{vx'}{c^2}\right)$$

$$x' + vt' = ct' + \frac{vx'}{c}$$

$$x' - \frac{vx'}{c} = ct' - vt'$$

$$x'\left(1 - \frac{v}{c}\right) = ct'\left(1 - \frac{v}{c}\right)$$

$$x' = ct'$$

and hence O' will also observe the speed of light to be c.

Comment: it was shown earlier in the book (see for example § 4.2.2 and § 4.2.3) that according to the velocity transformation (and composition) of Lorentz mechanics, the observed speed of light is invariant across inertial frames. In this exercise, the same conclusion is reached from the basic Lorentz spacetime coordinate transformations which the velocity transformations are derived from. This serves as a consistency check and confirmation to the validity of those results.

3. O and O' are two inertial observers in a state of standard setting where a light signal that propagates in all directions is emitted at $t = t' = 0$ at the common origin of coordinates. Using the invariance of the spacetime interval (which was established in § 3.9.4), show that if a 3D light signal is observed by O to propagate in a spherical shape[109] centered at his origin of coordinates, then O' should also see this light signal to propagate in a spherical shape centered at his origin of coordinates. Can you generalize the result? Comment on this question.

Answer: If O observed the signal to propagate in a spherical shape centered at his origin of coordinates, then this signal is described in O frame by the following equation:

$$x^2 + y^2 + z^2 - c^2t^2 = 0$$

This equation is no more than the equation of the spacetime interval in O frame. Now, since this interval is Lorentz invariant across all inertial frames then in O' frame we will have:

$$x'^2 + y'^2 + z'^2 - c^2t'^2 = 0$$

i.e. the signal will also be seen by O' to propagate in a spherical shape centered at his origin.

Regarding the generalization, we note that O' frame is an arbitrary inertial frame in a state of standard setting with O frame and hence the result is already general since it includes any frame in a state of standard setting with O. To generalize the result further by including all inertial frames (whether in a state of standard setting or not), we need to remove the condition "centered at his origin" and put the equation of sphere in a finite differential form and hence the signal is described in O frame by the equation:

$$(\Delta x)^2 + (\Delta y)^2 + (\Delta z)^2 - (c\Delta t)^2 = 0$$

---

[109] The spherical shape can be ascribed to the wave front of the signal.

9.3.2 Invariance of Observed Speed of Light

Hence, due to the invariance of the spacetime interval across all inertial frames, the equation of the signal in any other inertial frame will take the form:

$$(\Delta x')^2 + (\Delta y')^2 + (\Delta z')^2 - (c\Delta t')^2 = 0$$

which is also an equation of sphere.[110]

Comment: this question highlights the fact that having a constant observed speed of light across all inertial frames is equivalent to the invariance of spacetime interval across all inertial frames since they both originate from the same physical principle, i.e. the contraction of the c-calibrated spatial and temporal coordinates of spacetime by the same factor under the influence of motion which keeps the constancy of the observed speed of light across all inertial frames in an apparent sense. This fact will be more obvious if we put the above equations in the following form:

$$c = \sqrt{\frac{(\Delta x)^2 + (\Delta y)^2 + (\Delta z)^2}{(\Delta t)^2}} = \sqrt{\frac{(\Delta x')^2 + (\Delta y')^2 + (\Delta z')^2}{(\Delta t')^2}}$$

where the last equality arises from the fact that both the spatial coordinates in the numerator and the temporal coordinate in the denominator are scaled by the same γ factor which keeps the equality to c in both frames. So, if we assume the existence of an absolute frame then this speed is equal to c in the absolute frame in a "real" sense, while this speed is equal to c in all inertial frames which are in a state of motion relative to the absolute frame in an "apparent" sense (see § 1.6 and 11) due to the "scaling" of c itself in these frames because of the scaling of spacetime coordinates relative to the spacetime coordinates of absolute frame. In fact, both of these equivalent facts (i.e. having a constant observed speed of light across all inertial frames and the invariance of spacetime interval across all inertial frames) originate from the same principle which is using the speed of light in the calibration of spacetime measurements according to Lorentz mechanics.

4. Referring to the previous question, what about the source of light and if it should be in O frame or in O' frame for the concluded invariance to hold true?
   Answer: The state of the source of light (i.e. being in O frame or in O' frame) is irrelevant because although the velocity of light is frame dependent the speed of light is not, as we found in our analysis to the formalism of Lorentz mechanics (refer for example to the solved problems and exercises of § 4.2.2). To phrase the answer differently we would say: as indicated in the answers to previous problems and exercises, since the observed speed of the photon is assumed to be c in one of these frames (i.e. frame O according to the previous exercise), then any velocity component of this frame relative to the other frame (i.e. frame O' according to the previous exercise) should be passed to the photon as seen in the other frame according to our previous findings (i.e. the velocity of light should have a component of the velocity of its source although the

---
[110] That is: at any given instant of time, the wave front has a spatially spherical shape, i.e. $(\Delta x')^2 + (\Delta y')^2 + (\Delta z')^2 =$ constant.

9.3.3 Overall Assessment of Special Relativity Postulates 209

observed speed of light is invariant and hence it is independent of the velocity of its source). The essence of this answer is that although the frame in which the speed of light is observed to be c may not be physically the source of the light (i.e. it is not the rest frame of the light source), this frame is in lieu of a source frame since the speed of light takes its characteristic value c in this frame, and hence the observed speed of light should also be c in all other inertial frames which are in a state of uniform motion relative to this frame according to the Lorentz transformations (whether the velocity transformations or the spacetime transformations), as we found earlier (see for example § 4.2.2, § 4.2.3 and the present subsection). In brief, thanks to this invariance any inertial frame will be like a source frame for any light signal.

5. Compare between the representation of light signal in the ordinary 3D space and in the Minkowski 4D spacetime.
Answer: The light signal in the ordinary 3D space is represented by a sphere which represents the advancing wave front, while in the Minkowski 4D spacetime it is represented by a regular cone.[111] Both these forms of representation are invariant across inertial frames since both are based on the invariance of the spacetime interval and the invariance of the speed of light.

6. Discuss the claim that the second postulate of special relativity is confirmed by experimental observation.
Answer: It may be.[112] However, there is still space for questioning at least some of the claimed experimental evidence. For example, if we consider the Michelson-Morley experiment and its alike as some of these experimental verifications to the second postulate then all the arguments against the conclusions of this experiment (e.g. possible ether drag according to the wave theory, or potential projectile propagation model according to the ballistic theory) can invalidate this alleged evidence. Moreover, even if such evidence is well established, it cannot be used to support the special relativistic attachments to this postulate (e.g. considering the speed of light as restricted and ultimate in its wide sense) which some seem to consider as consequences or part of this postulate. Yes, we may consider the evidence in support of Lorentz transformations and Lorentz mechanics in general (see § 8) as indirect evidence for the invariance of the speed of light regardless of any interpretation (whether relativistic or not).

9.3.3 Overall Assessment of Special Relativity Postulates

1. Why we need to assume the existence of an absolute frame of reference?
Answer: We believe that the existence of an absolute frame is required for at least two reasons:
• To explain the real physical difference between inertial frames and non-inertial frames (see § 3.6.2).
• To comply with the principles of reality and truth (see § 1.6) by adopting a realistic

---

[111] "Cone" here has a more general sense which originates from its representation in a spacetime diagram with one temporal and two spatial dimensions.
[112] Regardless of the interpretation of the invariance and how it should be envisaged.

interpretation of Lorentz mechanics and its derived consequences and implications like length contraction and time dilation.

2. Based on the results that were obtained from analyzing the formalism of Lorentz mechanics, discuss the issues of light propagation model and the invariance of the speed of light and try to link these to the issue of absolute frame.
Answer: Referring to the analysis that we conducted previously (also see § 9.7.1), we can conclude that light behaves like projectile and hence its characteristic speed is relative to its source. This can be concluded from the added velocity component to the light signal from the velocity of its source. Accordingly, we do not need to assume a propagation medium (at least for this purpose) like ether to explain the propagation of light.[113] The invariance of the speed of light can then be explained by the contraction of spacetime coordinates by the motion through spacetime and for this purpose we need the assumption of an absolute frame to rationalize the obtained results. As we discussed before (refer to 1.14), the projectile model for the propagation of light is consistent with the existence of absolute frame.

In brief, there are two main elements that determine the propagation model of light in Lorentz mechanics: the variance of light velocity that is based on the added velocity component from the motion of the source which requires a projectile (instead of wave) propagation model, and the invariance of light speed which is based on the contraction of spacetime by motion through spacetime and this requires the existence of an absolute frame relative to which this real contraction occurs. In fact, the invariance of light speed is also partially justified by the added velocity component as indicated before and will be investigated further later on.

9.4 Abolishment of Fundamental Concepts

1. Contemplate on the abolishment of absolute space.
Answer: It is difficult to accept the abolishment of absolute space and admit the view that all motions are relative because it is inconsistent with the rules of physical reality and truth. Moreover, it is difficult to justify the difference between inertial and non-inertial frames without assuming an absolute rest space relative to which the frames are accelerating or not. Otherwise, why the space distinguishes between frames by treating some as inertial while treating others as non-inertial. The attempts to explain this by factors other than the existence of absolute space, such as explaining the difference by the existence of aggregates of matter and energy distributed throughout the space, do not provide a convincing answer because when we talk about the space we mean the physical space in which we exist not a hypothetical space which is devoid of matter and energy. Science is about discovering and understanding the real physical world that we live in where the space is one of the real aspects of this world. Yes, in philosophy or mathematics or in purely hypothetical contemplations we can imagine and characterize the space as we wish and derive the logical consequences of such a hypothetical space.

---

[113] In fact, we should not assume a propagation medium at least within a classical wave propagation model since this model is incompatible with a projectile propagation model.

9.5 Controversies within Special Relativity 211

So, in physics all laws and properties belong to the physical world and physical space including any real intrinsic attributes of this space such as containing matter and energy or having three dimensions. This should similarly apply to other attributes like time. In fact, the whole story of Lorentz mechanics and the justification of its emergence and rise is this desire to discover the real physical world; otherwise classical mechanics as a theoretical structure is completely consistent with the logic and we can keep it if we are not interested in the behavior and properties of the real physical world and the space of this physical world.

2. Discuss the following quote, which is attributed to Minkowski, and its significance in relation to the abolishment of space and time as separate entities: "From henceforth, space by itself, and time by itself, have vanished into the merest shadows and only a kind of blend of the two exists in its own right".

    Answer: This quote may be seen as a legitimate statement of the abolishment of space and time as separate entities and the emergence of spacetime as an established scientific fact. However, this should be seen within the framework of Lorentz mechanics but not beyond where space and time can still have independent existence.[114] This merge should also be seen within its formal context mainly as a mathematical artifact. Moreover, it is still partial and not thorough since we still see some distinction between spatial and temporal variables even within the framework of Lorentz mechanics. As we saw in several places in the text and exercises, the temporal and spatial coordinates of the spacetime are not equally treated in several aspects and hence the existence of a space which is separate and distinct from time is still there even in Lorentz mechanics. Anyway, even if we accept that space and time are totally merged and indistinguishable in Lorentz mechanics it does not mean that these concepts are abolished because they can still be used in other scientific and philosophical theories and they have full legitimacy to exist and represent real physical entities. We should also note that partial merge of space and time does exist even in classical mechanics where the Galilean transformation of space (i.e. in the x dimension according to the standard setting) also involves time although the merge in Lorentz mechanics is obviously more fundamental and extensive due to the effect of γ factor and the presence of spatial coordinates in the definition of frame time.

9.5 Controversies within Special Relativity

1. List some of the difficult questions that face special relativity.
    Answer: Some of these are:
    • Explaining the difference between inertial and non-inertial frames in the absence of absolute frame.
    • Deciding about contradictory views and opinions and flipflopping, e.g. if length contraction and time dilation effects are real or apparent and in what sense.
    • Refuting arguments like twin paradox (see § 10.1) where the relativity principle does

---

[114] As we indicated earlier, this is due to its philosophical and epistemological nature even though it has a basis in the formalism.

not apply equally and symmetrically to the involved reference frames.
- Assessing some thought experiments and abstract devices on which special relativity arguments and logic are based like the train thought experiment (see § 9.6.1) and the light clock (see § 9.7).
- Answering some challenges about the speed of light postulate such as claims of observation of superluminal speeds. In fact, some of these challenges represent a contest even to the current formalism of Lorentz mechanics and not only to the special relativistic interpretation.
- Deciding about the meaning of the constancy of the speed of light and adopting a clear interpretation.

9.6 Thought Experiments in Special Relativity

9.6.1 Train Thought Experiment

1. Describe another variant of the train thought experiment where a light signal is emitted at the center of the train.
   Answer: The relativity of simultaneity and the abolishment of absolute time may also be demonstrated in the literature of special relativity by a moving train with a light signal being emitted by a commuter at the center of the train in both directions at the time when the center of the train passes by a standing observer. Since the speed of light is the same for both observers and the train is moving towards the signal from the back and away from the signal in the front as seen by the stationary observer, while it is at rest for the on-board observer, the two events of meeting the front and back ends of the train will be simultaneous for the on-board observer and non-simultaneous for the stationary observer. This version of the train thought experiment (as phrased and presented in the circulating literature of special relativity) can also be challenged by some of the previous criticisms and disputes. For example, this version can be challenged by the dependence of the speed of light on the motion of its source and potential existence of an absolute frame as a medium for a classical wave propagation model. It can also be challenged by the difference between the simultaneity of occurrence and the simultaneity of observation. However, some of these challenges depend on the setting, assumptions, presentation and phrasing of this version of the train though experiment. Anyway, this does not affect the legitimacy and validity of these challenges because all these aspects are essential parts of any interpretative theory (especially when its legitimacy and logic are fundamentally based on thought experiments) and hence the experiment can be judged by these factors even if we assume that the experiment and its analysis can in principle be forged in a correct form. The reader is referred to the upcoming exercises of this subsection for formal analysis of two variants of the train thought experiment.
2. Discuss the difference between the simultaneity of observation and the simultaneity of occurrence.
   Answer: The simultaneity of observation means that two events are observed simultaneously by a particular observer in a particular position in a given frame, while the

9.6.1 Train Thought Experiment

simultaneity of occurrence means that two observed events occur at the same time in a given frame. For example, when I look to the sky and see the Sun and the Moon at the same time my observations are simultaneous. However, I know that these simultaneous observations do not come from simultaneous occurrences because the image of the Moon that I am seeing now represents the state of the Moon about 1.3 second ago while the image of the Sun that I am seeing now represents the state of the Sun about 500 seconds ago. Hence these simultaneous images are not produced by simultaneous occurrences (or events). Some of the special relativity arguments and thought experiments, including at least some variants of the train thought experiment, are based on the confusion between the simultaneity of observation and the simultaneity of occurrence where observation is treated as occurrence.

3. Make an argument in support of the claim that the train thought experiment is an example of the relativity of simultaneity of observation and not the relativity of simultaneity of occurrence.

   Answer: The whole experiment and its conclusions are based on the observation of localized observers and hence it belongs to the simultaneity of observation rather than the simultaneity of occurrence where the latter is based on the observation of a non-localized global observer in the given frame.

4. Regarding version 2 of the train thought experiment, justify why the added velocity component is rejected as a possible rationale for the special relativistic interpretation of non-simultaneity although this added component can be concluded from analyzing the formalism and hence it is accepted according to our view.

   Answer: As discussed in the text, it is logical to assume that if we adopt a projectile model for the propagation of light then the rest frame of the source of light in version 2 is the frame of the platform and hence the velocity of light should have a component from the velocity of its source as we found earlier. However, although this added component is logical and acceptable according to our interpretation since it does not affect the invariance of the speed of light where we considered the contraction of spacetime coordinates as the cause of this invariance and hence we accounted for this invariance despite the dependence of the velocity of light on the velocity of its source, in special relativity there is no such explanation; instead we find many explicit statements about the independence of the velocity of light from the velocity of its source although some (but not all) of these statements may be based on the tolerance and laxity of using "speed" and "velocity" interchangeably. Anyway, what is needed to rationalize the special relativistic interpretation is the variance of the speed of light, whether this variance is caused by the added velocity component or by something else, and this variance means the disposal of the second postulate of special relativity. Therefore, the presumed velocity component, whether accepted by special relativity or not, will not be able to rationalize the special relativistic interpretation of this version of thought experiment.

5. Analyze the two versions of the train thought experiment (as given in the main text) using the Lorentz spacetime coordinate transformations and hence conclude the relativity of simultaneity. Comment on the results.

9.6.1 Train Thought Experiment

Answer: It may be claimed that the logical thing to assume in this thought experiment is that the frame in which the two events are simultaneous is the frame of the source which should be the frame of train in version 1 and the frame of platform in version 2. However, we do not need to adopt a specific assumption or version to reach our conclusion about the relativity of simultaneity. So, let assume that in both versions the events are either simultaneous to P_1 (which we label as O) or simultaneous to P_2 (which we label as O' and we assume he is in a state of standard setting with O)[115] and hence we have two possibilities:

(a) The events are simultaneous to P_1: hence, we have $\Delta t = 0$ and we use the Lorentz time transformation from O to O', that is:

$$\Delta t' = \gamma \left(\Delta t - \frac{v \Delta x}{c^2} \right) = -\gamma \frac{v \Delta x}{c^2}$$

and hence $\Delta t' \neq 0$ because $\Delta x \neq 0$, i.e. the events are simultaneous for P_1 but non-simultaneous for P_2. It is obvious that $\gamma \neq 0$ and $v \neq 0$.

(a) The events are simultaneous to P_2: hence, we have $\Delta t' = 0$ and we use the Lorentz time transformation from O' to O, that is:

$$\Delta t = \gamma \left(\Delta t' + \frac{v \Delta x'}{c^2} \right) = \gamma \frac{v \Delta x'}{c^2}$$

and hence $\Delta t \neq 0$ because $\Delta x' \neq 0$, i.e. the events are simultaneous for P_2 but non-simultaneous for P_1.

Comment: using the formalism of Lorentz mechanics, we concluded the relativity of simultaneity easily and without confusion. Since, these transformations are based on (or the origin of) the contraction of spacetime coordinates due to the motion, the provided explanation should be sufficient and convincing. However, our interpretation of the relativity of simultaneity is different from the interpretation of special relativity because we believe that the relativity of simultaneity is consistent with the existence of absolute frame and hence absolute time. In contrast, the special relativity logic which is based on ill-stated and badly-presented arguments and thought experiments like this one can be challenged and hence the relativity of simultaneity may be rejected accordingly even though the result can be obtained correctly from analyzing the formalism. Moreover, the special relativistic interpretation of the relativity of simultaneity, which originates from (or the origin of) the denial of the existence of absolute frame and absolute time should be rejected due to the non-necessity of this denial for the explanation of the relativity of simultaneity (in addition to the need for absolute frame to distinguish inertial and non-inertial frames physically and realistically and to justify the reality of the characteristic Lorentzian effects like time dilation). We finally note that the above analysis shows that the events can be simultaneous to P_1 or to P_2 and hence

---

[115] We should remark that P_1 and P_2 in this analysis stand for the frames (i.e. they are global rather than local observers) since we are analyzing the thought experiment as an instance for the simultaneity of occurrence and not as an instance for the simultaneity of observation.

9.6.1 Train Thought Experiment

they are non-simultaneous to the other observer unlike the special relativity narrative where the events are simultaneous specifically to a particular observer. This in essence is based on the difference between our simultaneity (which is of occurrence) and the special relativity simultaneity (which should be of observation although it is meant to be of occurrence). In other words, our observers are global but the special relativity observers are local.

6. Analyze the other variant of the train thought experiment, which is given in exercise 1, using the principles of Lorentz mechanics as obtained from analyzing its formalism. Comment on the results.

   Answer: We consider the bystander and the commuter as two inertial observers, O and O' respectively, in a state of standard setting where a light signal that propagates in all directions is emitted at $t = t' = 0$ at the common origin of coordinates. We should also assume that the difference between the y and z coordinates of O and O' is negligible as if they are at the same point in space at $t = t' = 0$.[116] It was shown earlier (see the exercises of § 9.3.2) that using the invariance of the spacetime interval, if a 3D light signal is observed by O to propagate in a spherical shape centered at his origin of coordinates, then O' should also see this light signal to propagate in a spherical shape centered at his origin of coordinates. This means that O will see this signal as a spherical wave centered on his origin of coordinates, and similarly O' will see this signal as a spherical wave centered on his origin of coordinates. Now, we have two cases of simultaneity:

   (a) Simultaneity of arrival of light signal to the front and rear points of the platform: since O' is moving forward relative to the platform then his sphere (which moves with him) will hit the front point of the platform before it hits the rear point of the platform and hence these events are not simultaneous for O'. On the other hand, since O is standing still at the middle of the platform then his sphere will hit the front and rear points of the platform simultaneously. Accordingly, these events are simultaneous to O but non-simultaneous to O'.

   (b) Simultaneity of arrival of light signal to the front and rear points of the train: since O is moving backward relative to the train then his sphere will hit the rear point of the train before it hits the front point of the train and hence these events are not simultaneous for O. On the other hand, since O' is standing still at the middle of the train then his sphere will hit the front and rear points of the train simultaneously. Accordingly, these events are simultaneous to O' but non-simultaneous to O.

   Comment: we note that O and O' in our analysis are global (not local) observers and hence this is an instance of the relativity of simultaneity of occurrence rather than the relativity of simultaneity of observation. We should also remark that apart from the many reservations and question marks on the language and presentation of the special relativistic interpretation of this thought experiment which are found in the literature, an essential component in our interpretation is the fact that light has a

---

[116] In fact, this assumption is added for clarity and to match the narrative of this thought experiment; otherwise it is not needed because we are actually dealing with global observers representing frames of reference in a state of standard setting and hence the y and z coordinates should be identical.

velocity component from the velocity of its source (despite the invariance of its speed) which is essential to rationalize the interpretation of this thought experiment. Since there is no such velocity component in the special relativistic interpretation, then this experiment cannot be interpreted properly within the framework of special relativity.

9.7 Light Clock

1. Derive the time dilation formula from the light clock using a simple plot to illustrate the underlying physical principles.
   Answer: The physical principles of the light clock can be compactly presented in the illustration of Figure 11 and hence the derivation of the time dilation formula will be as follows:

$$\begin{aligned}
(c\Delta t')^2 &= (c\Delta t)^2 + (v\Delta t')^2 \\
(\Delta t')^2 &= (\Delta t)^2 + \beta^2 (\Delta t')^2 \\
(\Delta t')^2 - \beta^2 (\Delta t')^2 &= (\Delta t)^2 \\
(\Delta t')^2 (1 - \beta^2) &= (\Delta t)^2 \\
\Delta t' \sqrt{1 - \beta^2} &= \Delta t \\
\Delta t' &= \gamma \Delta t
\end{aligned}$$

which is the time dilation formula. We note that this formula may look different from a previous time dilation formula due to the exchange of prime. However, it is physically the same although it is notationally different. In fact, the assignment of prime to a particular frame is arbitrary and hence it has no physical significance on its own. What is significant is the actual physical setting. This should similarly apply to the arbitrary use of prime in other formulae which may also look different due to this notational artifact.

2. Show that light clock, as described in the text, follows the rules of classical mechanics in some aspects where the light signal behaves like a classical projectile.
   Answer: In Figure 12 we have an inertial observer O_1 who is on board a platform P that is uniformly moving to the right with respect to another inertial observer O_2. When O_1 throws a massive object (e.g. ball) upwards he observes his projectile following a straight line trajectory T_1 while O_2 observes this projectile following a parabolic trajectory T_2. The difference between the light clock and this example is the presumed presence of a gravitational field (which is pointing downward since the platform is supposed to be on the Earth) that makes the trajectory parabolic rather than straight. In the absence of such a gravitational field and the presence of a massive reflector (fixed to the platform) at the reflection point P_r the trajectory of the massive projectile according to O_2 will be made of the two straight segments (i.e. trajectory T_3) which is identical to the trajectory of the light signal in the light clock. The point of this question is that both the light signal (which can be represented by a photon) and the massive projectile have a sidewise velocity component and hence the light signal in the hypothesized light clock

9.7 Light Clock   217

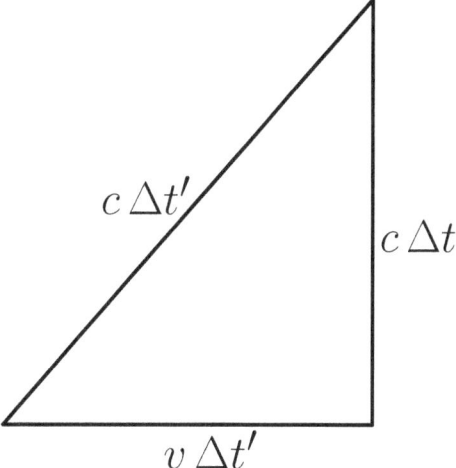

Figure 11: A simplified schematic diagram that illustrates the physical principles of light clock and the derivation of time dilation formula.

   follows the rules of classical mechanics in this regard. In other words, the light signal has a velocity component from the velocity of its source.
3. Make another argument for the case that the light clock follows the rules of classical mechanics in some aspects.
   Answer: The light path and the formulation for the observer who is in relative motion with respect to the clock is similar to those in the analysis of the Michelson-Morley experiment (refer to § 2.6 and § 12.2). It is obvious that the analysis of the Michelson-Morley experiment is based on a classical view. However, the Michelson-Morley analysis is based on a wave model while the light clock suggests a projectile model.
4. Assess a possible criticism to the light clock that its functionality (and hence time and time dilation) is based on its orientation.[117]
   Answer: Let assess the analysis of some special relativists to the light clock when it is rotated 90° so the light signal propagates in the same orientation as the velocity of the clock itself. According to this analysis, the time of the forward-backward pulse in the rest frame of the clock is:
   $$t_0 = \frac{2L_0}{c}$$
   where t_0 and L_0 are the proper time and proper length of the clock. Regarding the

---

[117] We note that there is nothing wrong in principle with the dependence of functionality on orientation since the physical principles on which the light clock rests can depend on orientation. The purpose of this question and its alike is to give typical examples about the framework of special relativity and its potential vulnerability to criticism due to questionable arguments and interpretations even when the results may be correct. In brief, our focus is potential criticism to this particular interpretation and its presentation.

9.7 Light Clock

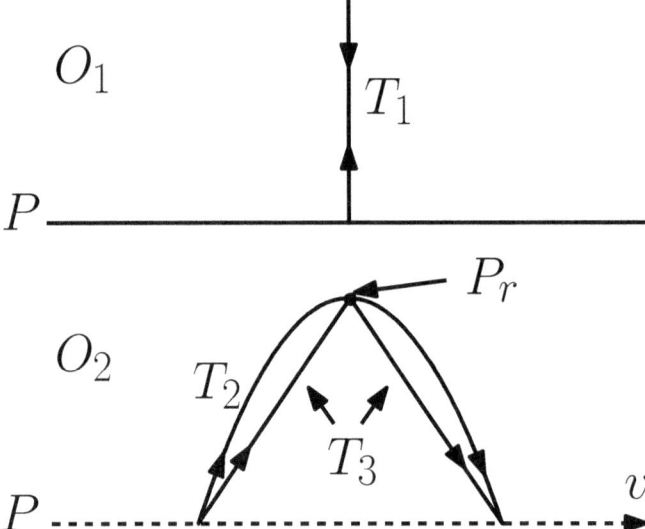

Figure 12: The trajectory of a massive projectile as seen by a moving observer O_1 who throws this projectile upwards (trajectory T_1) and as seen by a stationary observer O_2 in the presence of a gravitational field (trajectory T_2) and as seen by O_2 in the absence of gravitational field with the presence of a reflector at point P_r (trajectory T_3).

frame relative to which the clock is moving, we have:[118]

$$t = \frac{L}{c-v} + \frac{L}{c+v} = \frac{2Lc}{c^2 - v^2} = \frac{2L}{c(1 - v^2/c^2)} = \gamma^2 \frac{2L}{c}$$

where t and L are the time and length of the clock as measured from the moving frame and v is the relative speed between the two frames. Now, considering the time dilation effect which was derived from the light clock before its rotation (i.e. when the light signal was propagating in a perpendicular orientation to the velocity of the clock), we have:[119]

$$\begin{aligned} t &= \gamma t_0 \\ \gamma^2 \frac{2L}{c} &= \gamma \frac{2L_0}{c} \\ \gamma L &= L_0 \end{aligned}$$

---

[118] Note that in this analysis the special relativists overlook (deliberately or non-deliberately) the fact that the observed speed of light, $c \pm v$, as seen in the denominators is based on the Galilean velocity composition rule in utter disregard to the second postulate of special relativity. This should be supported by the fact that this analysis and formulation for the moving frame is identical to that of Michelson-Morley experiment for the light beam in the parallel arm (see § 12.2) which is obviously based on classical mechanics.

[119] We should remark that this analysis is based on assuming that the functionality of light clock is independent of its orientation (if the basis of time dilation is the functionality of the clock before its rotation) which is not obvious.

9.7 Light Clock

which is the length contraction effect. So, according to this analysis even length contraction effect can be derived from the light clock. However, this analysis should be challenged although the results may be claimed to be correct (see the analysis of this problem in § 9.7.1). For example, in the derivation of t the observed speed of light is assumed to depend on the speed of its source, i.e. the speed of light as measured from the frame which is in relative motion with the clock is assumed to be $c \pm v$ which is the classical formula for light speed. So, if we reject this classical speed because of its inconsistency with the second postulate of special relativity then we can argue (with no need for any formal derivation) that special relativists should abandon either length contraction effect or time dilation effect (or both) to keep their clock functioning as it is claimed.

Another (may be more fundamental) criticism to the light clock in special relativity is that the distinction between time and time measurement mechanism is rather vague in this theory where the common view is that the time measurement mechanism should follow the flow of time (see § 3.4) and hence if we accept that the functionality and time measurement mechanism of the light clock depend on its orientation then the flow of time itself in the frame relative to which the light clock is moving should depend on the orientation. More analysis to these issues will follow in § 9.7.1.

5. Make a formal argument for the case that if we accept the logic and argument of special relativity then the functionality and time measuring mechanism of light clock should depend on its orientation in a way that is inconsistent with the framework of this theory. Comment briefly on the result of your analysis.

Answer: As in the answer of the previous exercise, the time of the forward-backward trip in the rest frame of the clock following its rotation is:

$$t_0 = \frac{2L_0}{c}$$

Regarding the time of the forward-backward trip as measured from the moving frame, we have two possibilities:
(a) The observed speed of light follows the Galilean rule of velocity composition and hence we have:

$$t = \frac{L}{c-v} + \frac{L}{c+v} = \frac{2Lc}{c^2 - v^2} = \frac{2L}{c(1 - v^2/c^2)} = \frac{2L}{c}\gamma^2$$

So, if we assume time dilation we conclude length contraction and if we assume length contraction we conclude time dilation. However, this possibility is inconsistent with special relativity and its second postulate.
(b) The observed speed of light follows the second postulate of special relativity and hence we have (noting that d is the distance traveled by the clock during the time of

9.7 Light Clock

traversing L by the light signal):[120]

$$t = \frac{L+d}{c} + \frac{L-d}{c} = \frac{2L}{c}$$

which looks identical in form to the formula of the time as measured in the rest frame of the clock. So, if we assume that either time or length is identical in both frames then we can obtain neither time dilation nor length contraction. Otherwise, if we assume one of the two effects (i.e time dilation or length contraction) then the other effect will be obtained but in a sense that is inconsistent with the perspective and interpretation of special relativity.[121] Accordingly, accepting the logic of special relativity will lead to the conclusion that the functionality and time measuring mechanism of light clock depend on its orientation that is inconsistent with the framework of this theory and its postulates and principles. More details about these issues will follow.

Comment: in brief, if we accept the logic and arguments that we find in the literature of special relativity about light clock then we should either reject the second postulate of special relativity or reject one or both of the time dilation and length contraction effects, and this puts a question mark on the validity and logical consistency of the framework of special relativity as a legitimate interpretation to Lorentz mechanics.

6. Assess the consequences of the claim that the light clock is dependent in its functionality on its orientation within the framework of special relativity.
   Answer: The obvious consequence is that the time count and hence time dilation effect will be dependent on the orientation. Accordingly, time dilation will follow the style of length contraction in being orientation-dependent. This effect (i.e. dependence on orientation) is not supposed to occur to time. Anyway, the analysis in the answers of the previous questions suggests that if light clock should function according to the special relativity view in various orientations then we should abandon some of the special relativity principles and postulates.

7. What is the point of the previous exercises where some results that have already been shown to be obtainable from the formalism of Lorentz mechanics were challenged and shown to be wrong?
   Answer: The point is that the special relativity interpretation is questionable and hence the results are dubious if we adopt this interpretation. This means that these results, when they are correctly obtained from the formalism of Lorentz mechanics, indicate that the special relativity interpretation is not consistent with the formalism of Lorentz mechanics, and hence we should look for a logically consistent interpretation to Lorentz mechanics.

---

[120] This is inline with the analysis of light clock before its rotation where the Pythagoras theorem was used there to link the two components due to their orthogonal nature. Because the two components after rotation are along the same orientation a simple algebraic addition is used instead of the Pythagoras theorem.

[121] In fact, we may also need to assume that the functionality of the clock is independent of its orientation.

9.7.1 Assessing Light Clock

1. Repeat the analysis of the light clock for case (a) in the first question (refer to the solved problems) according to classical mechanics by using the Galilean velocity composition. Comment on the results.
   Answer: In classical mechanics we have a projectile propagation model and a wave propagation model and therefore we have two possibilities:
   (a) If we assume a projectile propagation model[122] for the velocity of light then the velocity of the light signal has a sidewise component v from the velocity of its source and a perpendicular velocity component c from its characteristic speed (which is relative to its source). Hence, the observed speed of light in O frame will be c while the observed speed of light in O' frame will be c' where:

$$c' = \sqrt{v^2 + c^2}$$

Accordingly, the time interval of a tick in O frame is:

$$\Delta t = \frac{L}{c}$$

while the time interval of a tick in O' frame is:[123]

$$\Delta t' = \frac{D}{c'} = \frac{\sqrt{d^2 + L^2}}{\sqrt{v^2 + c^2}} = \frac{\sqrt{(v\Delta t')^2 + (c\Delta t)^2}}{\sqrt{v^2 + c^2}}$$

Hence:

$$\Delta t' = \frac{\sqrt{(v\Delta t')^2 + (c\Delta t)^2}}{\sqrt{v^2 + c^2}}$$

$$(\Delta t')^2 = \frac{(v\Delta t')^2 + (c\Delta t)^2}{v^2 + c^2}$$

$$(\Delta t')^2 (v^2 + c^2) = (v\Delta t')^2 + (c\Delta t)^2$$

$$(v\Delta t')^2 + (c\Delta t')^2 = (v\Delta t')^2 + (c\Delta t)^2$$

$$(c\Delta t')^2 = (c\Delta t)^2$$

$$\Delta t' = \Delta t$$

which is consistent with the Galilean time transformation.
(b) If we assume a classical wave propagation model then the situation is more complicated because we have several possibilities with regard to the velocity of O or/and

---

[122] We may also call it ballistic propagation model.
[123] Although the previous formula may suggest that we are using c for spacetime calibration in our classical approach (as in the Lorentzian approach), the following formula should make it clear that this is not the case.

9.7.1 Assessing Light Clock

O' relative to the medium. However, we can consider one case in which O is at rest relative to the medium while O' is moving with respect to the medium with speed v in the opposite direction. The analysis and conclusions of this case are identical to those of the previous case (i.e. projectile model of part a) and hence we do not need to repeat. The analysis of the other cases will lead to the same results and conclusions where we will find that while the speed of light is frame dependent the time interval of a tick is frame independent.[124]

Comment: on comparing the classical results that we obtained in this question with the Lorentzian results that we obtained in the previous questions, we see (in consistence with our previous findings) that while in Lorentz mechanics the speed of light is invariant across inertial frames and the time is variant, in classical mechanics the opposite is true, i.e. the speed of light is variant across inertial frames and the time is invariant. As indicated earlier and will be seen in more details later, this is based on the difference in the methods used for the calibration of space and time in classical mechanics and in Lorentz mechanics.[125] Although, the calibration of space and time in classical mechanics may not be correct, the calibration of space and time (or spacetime) in Lorentz mechanics may not be the only correct (or the ideal) way of calibration and this should lead to the conclusion that a different theory (which is as good as Lorentz mechanics if not better) is possible in principle to emerge if we adopted a different method of calibration.

2. Show that the functionality of light clock is based on the fact that while the observed velocity of light is frame dependent, since it has a component from its source (and accordingly the characteristic speed of light c is relative to its source like classical projectile but with the added assumption of the contraction of spacetime coordinates under the influence of motion), the observed speed of light is frame independent.

   Answer: It was shown previously that the light signal in the light clock behaves like a classical projectile since its velocity has a component from the velocity of its source. However, it was also shown that the observed speed of light is equal to the characteristic speed of light c in both frames. Accordingly, while the observed velocity of the light signal is frame dependent, the observed speed of the light signal is frame independent.

---

[124] We note that our analysis here is different from the analysis of the Michelson-Morley experiment in the perpendicular arm (see § 12.2) because we are assuming that the light source is at rest with respect to the medium of propagation and the observer is moving, while in the Michelson-Morley experiment the light source is moving relative to the propagation medium and the observer is at rest with respect to the source.

[125] This is inline with our previous remarks that the most fundamental attribute in Lorentz mechanics is the speed of light, while the most fundamental attribute in classical mechanics is time and space. This is consistent with the fact that in Lorentz mechanics the speed of light is absolute while space and time are relative, but in classical mechanics the speed of light is relative while space and time are absolute. This is also inline with the fact that in Lorentz mechanics the invariant is the spacetime interval (which is based on the speed of light) while in classical mechanics the invariant is the time interval and space interval.

Chapter 10
Challenges, Criticisms and Controversies

10.1 Twin Paradox

1. Discuss the twin paradox and analyze its essence and implications.
   Answer: The twin paradox is based on the asymmetric time flow in two inertial frames whereas the relativity principle of special relativity is equally applicable in both frames due to the absence of absolute frame and hence each frame is referred in this "relativity" to the other frame rather than to a unique absolute frame. So, the essence of the twin paradox is the incompatibility between the principle of relativity in its special relativistic sense, where an absolute frame does not exist, and the existence of a privileged frame in which time runs slower or faster relative to the other frame. Hence, if we accept the restricted sense of the relativity principle, and hence we assume the existence of an absolute frame relative to which the time dilating motion takes place, then there should be no contradiction because the twin who is in a state of relative motion with respect to the absolute frame will experience time dilation (where time dilation is based on the γ factor of the motion relative to this absolute frame) whether he is the first or the second or both. But this challenge is valid if we embrace the special relativity interpretation because according to this theory everything is relative and hence time dilation effect should be symmetric with respect to the involved inertial frames and hence both frames should experience this effect as seen from the other frame. So in brief, the twin paradox is not applicable if we believe in the existence of a unique absolute frame, but it is a valid challenge if we believe in special relativity and its interpretation of the relativity principle which is based on the denial of the existence of absolute frame.
2. Discuss why a similar "twin paradox" is not usually proposed with respect to length contraction as proposed for time dilation.
   Answer: The reason is that in length contraction we should have the stick in the rest frame of only one of the two twins and hence the proper length belongs specifically to one frame. The situation of time is different because we have proper time in each frame since each observer has his own time. This may point out to another difference between spatial quantities and temporal quantities. However, we will see later (refer to § 10.2) that the barn-pole paradox can be regarded as the length contraction equivalent of the twin paradox. We may also synthesize a similar paradox where two identical sticks are placed in two inertial frames which are in a state of relative motion and hence each stick is shorter/longer than the other stick which leads to the violation of the principles of physical reality or to the rejection of the reality of length contraction. Alternatively, we may propose a similar paradox where a stick in a third frame C is observed from two inertial frames (A and B) which are in a state of motion relative to each other and relative to C where the supposed length contraction will be different according to the frame of observation, i.e. A or B.

3. Discuss the logical inconsistency between the first postulate of special relativity and the twin paradox.
   Answer: The twin paradox puts a question mark on the validity of the principle of relativity (i.e. the first postulate) in its special relativistic sense with regard to the aging of the traveling twin because there is a real physical effect (i.e. slow aging of the traveling twin) which is a function of the speed of the frame and this real effect cannot be explained in the absence of absolute frame. Hence, the reality of slow aging and the denial of absolute frame are logically inconsistent since the former requires an asymmetric situation between the two twins while the latter implies a completely symmetric situation between the two twins.

10.1.1 Time Dilation Effect is Apparent

10.1.2 Traveling Twin is Distinguished by being non-Inertial

1. Analyze the traveling twin motion as partly inertial and partly accelerating, and hence derive the logical consequences of this analysis.
   Answer: We may contemplate that the journey of the presumed traveler who accelerates and decelerates, as well as uniformly moves as an inertial observer in part of his journey, should be split into inertial part where the twin paradox should be addressed since this part is subject to Lorentz mechanics, and non-inertial part which should be subject to the analysis of accelerated motion. So, if we assume that accelerated motion relative to an inertial observer is subject to Lorentz mechanics (by considering the time dilation as applied to the instantaneous inertial frames for instance), then the time required for the journey should be the sum of the inertial time and non-inertial time where the non-inertial time is obtained by integration. This analysis should lead to the conclusion that this answer (i.e. traveling twin is distinguished by being non-inertial) will not solve the problem (at least in the inertial stages where time dilation should apply symmetrically to both twins) and hence the twin paradox is valid if we adopt the special relativity interpretation.[126]

10.1.3 Traveling Twin has Two Inertial Frames

1. Assess the concept of "change of frame" in the context of twin paradox.
   Answer: We label the twins as A and B. In the absence of absolute frame we can say:
   • From the perspective of A frame (or any other frame which is at rest with A frame) B changed his frame.
   • From the perspective of B frame (or any other frame which is at rest with B frame)

---

[126] We do not go through potential involvement of general relativity in the solution of the non-inertial part because this is out of the scope of the book. Moreover, ascribing the acceleration to a particular frame in the absence of an absolute frame still requires an answer since the symmetry is still there whether we are in the domain of special relativity or general relativity. In fact, general relativity is like special relativity in its need (to be sensible and logically consistent) for an absolute frame at least in dealing with accelerating frames because general relativity has no magical power to define acceleration realistically with no reference to an absolute frame.

A changed his frame.
- From the perspective of any other frame C, one or both changed frame (depending on C).

10.1.4 Calling for General Relativity

10.2 Barn-Pole Paradox

1. Discuss the relation between the relativity principle and the barn-pole paradox.
   Answer: This paradox arises if we accept the relativity principle in its special relativistic sense where length contraction effect is symmetric or reciprocal and hence it applies to both frames due to the absence of absolute frame. Hence, this paradox has no place if we adopt an interpretation that is not based on the relativity principle of special relativity. For example, if we accept the restricted form of the relativity principle and hence length contraction occurs only to the object that moves relative to the absolute frame then the pole will fit if it is the object that moves relative to the absolute frame because it is the contracted object, and the pole will not fit if the barn is the moving object because it is the contracted object. In the case that both pole and barn move relative to the absolute frame, the pole may fit inside the barn or may not fit depending on the amount of contraction that each one has suffered. For example, if a 2 m pole contracted to 1 m and a 1.5 barn contracted to 1.2 m due to their motion relative to the absolute frame then the pole will fit inside, while if the 2 m pole contracted to 1.5 m and the 1.5 barn contracted to 1.3 m then the pole will not fit.

2. A 2 m long pole is required to fit inside a 1 m long barn. What is the minimum speed required to do this fitting? In which frame (barn frame or pole frame) this fitting can happen? What about the other frame involved in this process? Base your answer on special relativity.
   Answer: If a 2 m long pole is to fit inside a 1 m long barn then the pole length should be contracted to at least 1 m. From the length contraction formula we have:

$$L_{pc} = \frac{L_p}{\gamma}$$

where L_p is the pole proper length and L_{pc} is its contracted length. Hence:

$$\gamma = \frac{1}{\sqrt{1-\beta^2}} = \frac{L_p}{L_{pc}} = \frac{2}{1}$$

$$\sqrt{1-\beta^2} = \frac{1}{2}$$

$$\beta^2 = \frac{3}{4}$$

$$\beta = \frac{\sqrt{3}}{2}$$

$$v = \frac{\sqrt{3}}{2}c$$

which is the minimum speed required to do the fitting. This fitting is possible according to the barn frame where the pole is seen moving and hence it is contracted. From the pole frame, the barn will contract according to the length contraction formula, that is:

$$\begin{aligned} L_{bc} &= \frac{L_b}{\gamma} \\ &= 1 \times \sqrt{1 - \beta^2} \\ &= \sqrt{1 - 0.75} \\ &= 0.5 \, \text{m} \end{aligned}$$

where L_b is the barn proper length and L_{bc} is its contracted length. Hence, the situation will be worsened because we then should try to fit a 2 m long pole into a 0.5 m long barn which is more difficult than fitting a 2 m long pole into a 1 m long barn.

3. Discuss the special relativistic answer to the barn-pole paradox challenge that the fit will occur but because of the limited speed of sending signal from the front end of the pole to its back end the problem will occur later.

Answer: This special relativistic answer is based on assuming that the pole (rather than the barn) will contract and the fit will happen but the problem will arise after that when the pole stops following the hit. Regardless of a number of potential challenges and question marks, this answer fails to address the essence of this paradox which is the incompatibility between the relativity principle in its special relativistic sense (i.e. in its general applicability and hence the effects should be symmetric since the principle equally applies to both frames) and the asymmetric nature of length contraction effect that presumably happens to the pole, rather than the barn, according to this answer. Interestingly, some special relativists tried to improve this answer by assuming that the front end will stop as soon as it hits the front wall of the barn but because the speed of sending signals is finite the back end (as well as the other parts which are still unaware of the hit) will continue its journey towards the inside of the barn until it enters the barn and fits inside where the problem occurs. Apart from the fact that this answer does not address the problem of the logical inconsistency which is based on the incompatibility between the principle of relativity in its special relativistic sense and the asymmetric nature of length contraction, we need to know if the speed of sending signals inside the pole (which should depend on the nature of the pole material) is sufficient to make this fit possible, at least partially where the rear end and its neighborhood may have sufficient time to continue their journey assuming that the stop of the in-between parts will not hinder this. Moreover, this answer will change the nature of the problem from being a length contraction issue, which is a logical problem related to the nature of space and time, to a material science problem because the argument unwittingly assumes that length contraction did not happen during the movement of the pole before it hits the front wall but it happened after the hit due to the presumed compression which occurred since the rear parts will continue their journey thanks to their ignorance of the hit. We should also remark that this answer may need to assume that the pole parts are independent of each other in their movement as if the internal forces will not arise until the signal is received. A perfectly elastic spring in place of a rigid pole may be more appropriate to assume to make some sense of this answer. In brief, this answer is totally worthless because it is based on complete misunderstanding of the nature of length contraction effect and the essence of this paradox since length contraction, according to this answer, is caused by mechanical compression of the pole (rather than by spacetime coordinate transformations) due to the hit and hence it does not exist before hitting the front wall.[127] We should finally remark that according to the logic of this answer the fit should depend on the amount of compression, which should depend on the nature of the pole and its speed, and hence it is not subject to the length contraction formula.

4. What is the significance of the barn-pole paradox according to special relativity?
   Answer: The significance is that it highlights the fact that order relations (i.e. smaller than, equal, and greater than) are not well defined in a real sense if we accept the special relativistic interpretation of Lorentz mechanics.

10.3 Other Paradoxes

10.4 Speeds Exceeding c

10.5 Non-Local Reality of Quantum Mechanics

---

[127] The strangest thing about this answer is that it assumes that length contraction does not occur during the state of motion but during the state of rest; opposite to what it should be.

Chapter 11
Interpretation of Lorentz Mechanics

1. What is the relationship between the experimental evidence of a theory and its interpretation?
   Answer: The relationship between the experimental evidence of a theory and its interpretation is an epistemological issue. The interpretation is essentially a framework that is based on logical, epistemological and philosophical premises whose purpose is to rationalize the formalism of a certain physical theory. This framework may also contain some premises of scientific nature. We may contemplate that if this framework is the only possible interpretation and explanation of the formalism, then any evidence in support of the formalism may be regarded as evidence in support of this framework and hence this framework may get the legitimacy and endorsement of the experimental evidence like the formalism itself. However, it is virtually impossible to rule out other potential interpretations that could emerge in the future and can be similarly valid in explaining and interpreting the formalism even if we currently see only one possible interpretation. Hence, it is difficult to regard the scientific evidence as an endorsement to any particular interpretation. Regarding the scientific elements in the framework, if the evidence that supports the formalism can be seen as evidence for a particular element, it can be regarded as evidence in support of that element; otherwise it can not.

2. Refute the claim that the validity and invalidity of interpretation is of little practical value to science since in the presence of a correct formalism the interpretation will have little impact on the progress of science even if the interpretation was wrong.
   Answer: This is not true in general because the interpretation can be as important as the formalism itself in steering the scientific research and providing a sense of direction and a compass for the future development. So, a wrong interpretation of a correct formalism can be as disastrous as a wrong formalism since it can result in hindering the research in certain areas and directions and motivating the research and directing the resources in the wrong direction. The obvious example for this in our view is special relativity where the wrong interpretation, despite the "correct" formalism, resulted in a number of tragedies in modern physics.

3. Outline the main features of the Lorentz interpretation of Lorentz mechanics and compare it to the interpretation of special relativity.
   Answer: Based on the available records, the main features of the Lorentz interpretation are summarized in the following points:
   • Space and time are absolute and hence there is an absolute frame. In special relativity, space and time are relative and hence there is no absolute frame.
   • Simultaneity is absolute. In special relativity, simultaneity is relative.
   • The ether does exist. In special relativity, ether does not exist or at least it is superfluous.

- Effects like length contraction and time dilation are real. In special relativity, effects like length contraction and time dilation are real in one opinion and apparent in another opinion.
- The invariance of the speed of light across inertial frames is apparent. In special relativity, there is confusion and contradiction about this issue although the general stand seems to adopt the real option.

11.1 Criteria for Acceptable Interpretation

1. Try to justify the above criteria for accepting any proposed interpretation of Lorentz mechanics.
   Answer: The justification of these criteria should be obvious as explained in the following points:
   - The interpretation should be logically consistent because the science that is not based on the fundamental rules of logic is nonsense and hence it may be useful for any purpose but not for building sensible and reliable understanding of the physical world because no understanding can be built without the rules of logic.
   - The interpretation should be compliant with the principles of physical reality and truth because the science that is not based on these rules is no more than fiction and illusion and hence it does not deserve any attention.
   - The interpretation should be consistent with other known facts and well-established theories because any scientific theory that is not consistent with other facts and established theories should be wrong according to the principles of reality and truth. Although there are different branches of science and human knowledge in general, they should all reflect the same reality and hence we should have a single truth that represents the unique physical reality. This, in fact, represents the necessity for the unity and conformity of science and knowledge in general.
   - The interpretation should be thorough because what is required is an interpretation to Lorentz mechanics and not to certain elements of this mechanics. Partial interpretations usually lead to contradictions and inconsistencies where they may succeed in explaining certain elements and issues but they clash with other facts. In fact, partial interpretations should be discarded by default because their partial and normally *ad hoc* nature is usually based on their lack of correct and sufficiently general framework to explain the theory as a whole; otherwise they should be complete and general.
2. What is the relation between science and logic?
   Answer: Compliance with logic is a necessary but not sufficient condition for scientific theories and hence any scientific theory that is not consistent with the rules of logic should be rejected although not every scientific theory that is consistent with the rules of logic should be accepted since its acceptance also requires the support of experimental and observational evidence. For example, classical mechanics as a theoretical structure is totally consistent with the rules of logic but it is not necessarily true as a scientific theory because the laws of classical mechanics are not completely consistent with our observations of the physical world.

11.1 Criteria for Acceptable Interpretation

3. Give some examples of attempts made by special relativists to defend the logical inconsistencies of special relativity.
   Answer: Examples are:
   • Some special relativists tried to confuse the issue by identifying the logic with common sense which generally, but not necessarily, based on logic.[128] Although scientific theories are not required to comply with the rules of common sense as such, they are required to obey the rules of logic.[129]
   • Some special relativists tried to defend special relativity and its logical inconsistencies by questioning the authority of the logic itself and casting a shadow on its validity to judge scientific theories as if science is not required to be subject to the logic and compliant with its rigorous rules. In fact, some special relativists even ridiculed the logic itself as if it is a collection of old fictitious beliefs that should be rejected by modern physics, while others went further by considering "throwing the *old logic* in the bin" as one of the great achievements of special relativity. Throwing the logic in the bin means living in the kingdom of nonsense where not only science does not exist but even human in his intellectual existence as we know will cease to exist.
   In brief, science with no logic is no more than nonsense. So, all these lines of defence are not worth any effort to investigate or refute.

4. Discuss the stand of some special relativists who tried to challenge the logic by experimental evidence in support of special relativity and its postulates.
   Answer: This stand can be refuted by the following:
   • There is no evidence in support of special relativity and its postulates as such. All the true and alleged experimental evidence in this regard are in support of the bare formalism of Lorentz mechanics and not of any particular interpretation and its epistemological and philosophical foundations, such as special relativity and its postulates. This is like the situation in quantum mechanics, where the evidence in support of quantum mechanics is generally seen as evidence for the formalism and not for any particular interpretation such as the Copenhagen school or the parallel worlds interpretation.
   • No experimental evidence can challenge the logic. In fact, no correct experimental evidence can contradict the logic because no reality or truth can contradict the logic. Any experimental evidence should be processed, rationalized and judged by the logic not the other way around. Hence, we cannot reject or modify the rules of logic by alleged experimental evidence. That is how human mind evolved and how it works. Logic represents the most fundamental principles of our perception system and the blueprint of all our mental processes. Logic is the infrastructure for all our intellectual achievements including not only science, but even mathematics. Accordingly, no experimental

---

[128] We may define common sense in this context as generally accepted rules that are usually, but not necessarily, based on the rules of logic and daily life observations. For example, "the impossibility of having something and its opposite at the same time and from the same perspective (e.g. I exist now and I do not exist now)" is a rule of logic, but "the weak does not conquer the strong" is a common sense rule because there is no logical contradiction in conquering the strong by the weak although this rule is generally true from our daily life experience and it does not contain any logical inconsistency.

[129] An example of an acceptable scientific theory that may not be fully compliant with common sense is quantum mechanics.

11.2 Essential Elements of Potentially Acceptable Interpretation

evidence can rectify or justify the faulty logic of a theory.

11.2 Essential Elements of Potentially Acceptable Interpretation

1. Explain why the existence of an absolute frame is needed to have a valid interpretation.
   Answer: The existence of an absolute frame is needed to justify the difference between inertial and non-inertial frames and to rationalize the reality of characteristic Lorentzian effects like time dilation and length contraction where the reality of Lorentz mechanics and its consequences is dependent on this reality. Hence, the existence of such a frame should be assumed and the relativity principle should therefore be interpreted in its classical sense (see § 2.2).

2. Try to explain why inertial frames are treated equally by the space in their validity for formulating the known laws of physics despite their possible motion with various speeds relative to the absolute frame while non-inertial frames are treated differently.
   Answer: We may propose that all inertial frames are equal because they are not accelerating relative to the absolute space and hence they can all be considered as being at rest relative to this space due to its infinite extension, while non-inertial frames are different because they are accelerating relative to absolute space. However, the difference between inertial and non-inertial frames can be explained more fundamentally by the fact that velocity is a first order variation in spacetime while acceleration is a second order variation. The difference that is represented by the spacetime contraction between the inertial frames which are at rest with respect to the absolute frame and those which are in a state of uniform motion with respect to the absolute frame may then be explained as a first order variation effect while the difference between inertial frames and accelerating frames can be explained as a second order variation effect. In brief, the inertial frames are differentiated from each other by a first order variation in spacetime while the inertial frames are differentiated from non-inertial frames by a second order variation in spacetime. The known laws of physics should then be assumed to be independent of the first order variations and dependent on the second order variations. If this is established, then we may expect other differentiations between frames by higher order variations in spacetime where in these differentiations even non-inertial frames will be categorized according to the order of variation and the type of dependency of the laws on the order of variation.[130]

3. Demonstrate the invariance of the speed of light and the non-invariance of its velocity using a simple diagram.
   Answer: The diagram should look like Figure 9, where the invariance of the speed of light is represented by the fact that the vectors AB and AB' (which represent the velocity of light in frames O and O' respectively) have the same magnitude since these

---

[130] Some aspects about the rules of accelerating frames and the laws in these frames may be referred to general relativity. However, the link between special and general relativity is a controversial issue. The geometry of spacetime which may be used to provide the link and hence solve some problems of special relativity may be contested. These issues require deep inspection and analysis which we intend to address in the future.

vectors are radii of the shown circle, while the non-invariance of velocity is represented by the fact that these vectors have different directions in space since they have different x and y velocity components.

4. Why the formalism of Lorentz mechanics indicates a projectile-like model for the propagation of light?
   Answer: The velocity component of light signal from the velocity of its source suggests that light follows a model of propagation similar to the classical model of projectile (see § 1.13.1 and § 1.14) although the speed does not follow this classical model because of the Lorentzian spacetime contraction which does not exist in classical mechanics.[131] Hence, if we follow the formalism of Lorentz mechanics then the characteristic speed of light c should be relative to its source like classical projectile but with the added assumption of the contraction of spacetime coordinates under the influence of motion. As we discussed earlier, the projectile model of propagation is logically consistent with the existence of absolute frame although it (unlike the classical model of wave propagation) does not require the existence of such a frame.

5. Outline a potential interpretation that naturally emerges from analyzing the formalism of Lorentz mechanics and incorporates the proposed elements in this section.
   Answer: We may propose such an interpretation from our previous analysis and findings. This interpretation, whose origin can be traced back to our investigation in § 1.6, is outlined in the following points:
   • Space and time are absolute. Hence, we have an absolute frame of reference relative to which observers and frames can be at rest or in uniform motion or in accelerated motion. For convenience, we call an observer who is at rest with respect to this absolute frame "absolute observer". Similarly, we call an observer who is in a state of relative motion with respect to this absolute frame "relative observer" and we call his frame "relative frame". We note that in essence we have only one absolute frame because since all absolute frames should be at rest relative to each other, their coordinate systems can be reduced to a single coordinate system by simple static transformations, i.e. translation, rotation, reflection and scaling of axes. Moreover, since their time is absolute, their temporal scales can be unified by simple translation and scaling to unify the origin of time and its unit interval. Similarly, we can reduce the infinitely-many absolute observers to a single absolute observer who may be called *the* absolute observer.
   • Let define the reality as the state of the objects as observed by the absolute observer and the truth as his observation. Hence, all other states (or realities) and observations (or truths) are "apparent" in the sense that they are obtained under the influence of length contraction and time dilation, as will be discussed next. It may be claimed that this is just a useful convention to avoid having multiple realities and truths. Anyway, we do not mind if someone decided to consider all these observations as "real" as long as there is no difference in the interpretation since this will not lead to any violation to the principles of reality and truth due to the difference in perspective.
   • Length contraction and time dilation are real fundamental physical effects that in

---
[131] We note that the Lorentzian spacetime contraction should be linked to the use of the speed of light for calibration of spacetime measurement as explained earlier.

11.2 Essential Elements of Potentially Acceptable Interpretation   233

essence represent the same phenomenon which is the shrinking of spacetime coordinates under the influence of motion relative to the absolute frame where length contraction is the spatial demonstration of this phenomenon while time dilation is the temporal demonstration of this phenomenon. It was shown earlier (refer to § 5.1.5) that length contraction and time dilation effects represent observations from different perspectives. We may therefore unify their perspectives and hence call both these effects "contraction" (i.e. length contraction and time contraction) although we may also call both these effects "dilation" (i.e. length dilation and time dilation) from the opposite perspective. We may even enhance this unification by calling these effects "contraction of spacetime". As well as being useful for reducing the confusion, this will also grant the temporal and spatial coordinates similar status and hence make the coordinates of spacetime more homogeneous and unified.

• Since spacetime contraction is real physical effect, then the explanation of the measurement of different length and time in the absolute and relative frames is that since the length and time measuring mechanisms in the relative frames will also suffer from the same length and time contraction that affects the measured objects in these frames, then the observers in the relative frames will not see the effect of this length and time contraction, i.e. they will have the same measurement as if they were at rest relative to the absolute frame.[132] So, as the speed of the relative observers with respect to the absolute frame changes, both the measuring mechanisms and the measured objects will contract by the same factor and hence they will not observe any difference and they will always obtain the "proper" values from their measurements.

• The explanation that is given in the previous points about length and time contraction will solve the problem of reality and truth because all observers will deal with some sort of reality and truth due to the fact that the "apparent" observations in the moving frames are not so because they are illusions and have no physical basis but because they are observed by using contracted length and time measuring mechanisms so all observers observe some sort of reality and have some version of truth, i.e. their own reality and their own truth which in a sense are identical to the reality and truth that they will have if they are in the absolute frame. However, the ultimate reality and truth belong to the absolute frame. In other words, because the measurements of the relative observers are different from the reality of absolute frame, these measurements may be described loosely as "apparent" but this is totally different from the meaning of "apparent" as equivalent to illusion.

• Inertial frames are those frames which are at rest or in a state of uniform motion with respect to the absolute frame, while non-inertial frames are those frames which are in a state of accelerated motion with respect to the absolute frame. Moreover, the absolute frame does not distinguish between inertial frames in their validity for formulating the known laws of physics although it distinguishes between inertial and non-inertial frames (and possibly between non-inertial frames themselves but this is outside the scope of Lorentz mechanics which is the subject of our investigation). So, the known laws of

---

[132] This justifies the fact that the proper value is the same in all frames despite the potential contraction of spacetime.

11.2 Essential Elements of Potentially Acceptable Interpretation

physics equally apply to all inertial frames but not necessarily to non-inertial frames.

• Because of the existence of absolute frame, the relativity principle applies in its classical sense but not in its special relativistic sense. Accordingly, all the formulations of Lorentz mechanics should be ultimately referred to the absolute frame. In particular, the speed of relative motion v in the Lorentz γ factor and other formulae will be ultimately referred to the absolute frame although in some formulations (e.g. kinetic energy) they may be referred to relative frames.

• Based on the previous points, the observations of all relative observers who are in a state of relative motion with respect to each other will be naturally calibrated by the observations of the absolute observer. For example, let have A as the absolute observer and B and C as two relative observers who are in a state of relative motion with respect to each other. Now, if A measures a given stick that is at rest in the B frame to be 0.5 m long and B measures this stick to be 1 m long then the stick is contracted by a factor of 0.5 in the B frame due to its relative motion with respect to the absolute frame. Similarly, if A measures another stick that is at rest in the C frame to be 0.25 m long and C measures this stick to be 1 m long then the stick is contracted by a factor of 0.25 in the C frame due to its relative motion with respect to the absolute frame. Accordingly, B will measure the C stick to be 0.5 m long while C will measure the B stick to be 2 m long. It should be obvious that when all these observers gather in a single frame (whether in the absolute frame or in a relative frame) they should agree on all measurements and hence both these sticks will measure 1 m long by any one of these observers (who become essentially a single observer due to their gathering in a single frame). The calibration in the case of having different proper length of the B and C sticks should be straightforward.

• Since time is absolute, simultaneity is absolute as it refers to the time of the absolute frame. Accordingly, the relativity of simultaneity in its special relativistic sense should be abandoned although relative observers may disagree on the simultaneity of events or their chronological order.[133] However, they should agree when they refer their events to the absolute frame by standardizing their time to the absolute time. The same can be said about the relativity of co-positionality in its relation to absolute space.

• Regarding the speed of light, if we accept that the real value of the speed of light c belongs to the absolute frame, but at the same time all other observers will have an identical apparent measurement of the speed of light as c because their length and time measuring mechanisms will scale by the same Lorentz γ factor, then the issue of the invariance of the speed of light is solved since all observers will measure this speed to be c although this measurement is real only for the absolute frame while it is apparent for all other frames.[134] Regarding the speed of the light source, it has no impact on the speed of the emitted signal because the added velocity component to the signal from

---

[133] This may be likened to the relativity of simultaneity of observation despite the fundamental difference between the two.

[134] In fact, c is c in any frame, so what we mean by having real and apparent speed of light is that this speed is a ratio of original (real) or contracted (apparent) spacetime coordinates where in the latter case it can be considered as a scaled version of the real c with a scale factor of unity.

11.2 Essential Elements of Potentially Acceptable Interpretation

the velocity of the source is neutralized by the effect of displacement of objects in the observed frame (refer to § 9.7.1). In brief, the speed of light as referred to any frame (whether the observer frame or another frame) will be the constant c because the units of time and space measurements in that frame are scaled by the same factor as the time and space in that frame while the velocity of the source in that frame will have no effect because this velocity will be added to the velocity of light in that frame and this added velocity component will be annulled by the effect of displacement of objects in that frame where this displacement is due to the frame motion.[135] At the root of all this is the choice of c as a standard for calibrating the measurements of spacetime.

• Regarding the issue of the speed of light as restricted and ultimate speed (see § 3.8.1), we repeat what we stated previously, i.e. this should be considered within the validity domain of Lorentz mechanics and according to its current formalism. Moreover, the difference between massive and massless objects should be considered. Hence, we believe that the taboos of special relativity about the speed of light should be re-assessed and may be abolished as some of them lack any basis from the analysis of formalism and any experimental evidence.

• All other elements and characteristic features of Lorentz mechanics, such as the merge of space and time into spacetime and the use of the speed of light as standard for calibrating space and time measurements, are consistent with the proposed interpretation as they enter in the formal structure of Lorentz mechanics and hence they provide the basis for this interpretation.

6. Inspect and analyze the logical and physical consequences of the interpretation that was proposed in the previous question to assess its validity.
   Answer: Thorough inspection and analysis requires very detailed discussions and deliberations which are beyond the capacity of this exercise and the intended size and objective of the book. However, we may outline our analysis and assessment by a few examples in the following points:
   • The difference between inertial and non-inertial frames will be naturally explained by the proposed interpretation, that is the absolute frame does not discriminate against the frames which are in a state of uniform motion with respect to the absolute frame but the absolute frame does discriminate against the frames which are in accelerated motion with respect to the absolute frame.[136]

---

[135] In other words, when we measure the speed of light in reference to a particular frame we should use the length and time measuring equipment and units of that frame, and as soon as we do so the effect of the added velocity component will be annulled and what remains is the scaled units of length and time (due to spacetime contraction) which ensure the invariance of light speed.

[136] As indicated earlier, this discrimination is based on the order of variation of spacetime on which the laws are based and hence we need a more fundamental definition of the "known laws of physics" whose form does not vary between different inertial frames. In brief, the laws of physics should not vary between frames if they do not depend on a certain order of spacetime variation that distinguishes these frames but they should vary between frames if they do depend on that order of spacetime variation. This proposed principle requires further investigation and development. Anyway, this puts the existence of absolute frame on a more firm foundation since the reality of these variations as the bases for the difference between frames is totally dependent on the reality of the absolute frame since

11.2 Essential Elements of Potentially Acceptable Interpretation

- The length contraction and time contraction are real physical effects as explained above. Now, since the relativity principle applies in its classical sense due to the existence of absolute frame, challenges like twin paradox and barn-pole paradox will be naturally answered because these effects do not apply equally and symmetrically to all frames, i.e. they apply only to relative frames and observers but not to the absolute frame and observer and hence any contraction in length or time in any frame will be uniquely and unambiguously determined by the motion of the frame relative to the absolute frame.[137]
- The invariance of the speed of light across inertial frames will find a simple explanation that is any observer will obtain the proper values of the objects in his frame from his measurements despite the contraction of spacetime. Now, since light belongs to each frame equally (where the added velocity component from the velocity of the source will compensate for the motion of the frame) then the "proper" (or characteristic) value of its speed (i.e. c) should be equally obtained in any frame.

7. Explain length contraction and time dilation effects from the viewpoint of the proposed interpretation of Lorentz mechanics and outline their relation to the absolute frame and to relative frames.
   Answer: Space and time contract in any frame which is in relative motion with respect to the absolute frame. This contraction is the basis for the length contraction and time dilation effects. Accordingly, these effects are real since they are observed from the absolute frame. Furthermore, the proper length and time as measured in any other frame (i.e. frames moving with respect to the absolute frame) are apparent because the measuring rods and time measuring mechanisms will suffer the same contraction that the measured length and time suffer. In brief, length contraction and time dilation are real but the measured values of lengths and time periods in the relative frames are apparent.

8. Imagine two inertial observers in a state of standard setting. Following the instant $t = t' = 0$, each one of these observers will keep watching the time of the other observer. What will they see? Base your answer on the proposed interpretation.
   Answer: This depends on their relative motion with respect to the absolute frame and hence one observer can see the time of the other observer lagging behind his time while the other observer sees the opposite. They can also see their times totally synchronized.

9. Our observations, such as the prolongation of lifetime of elementary particles, should suggest, according to the proposed interpretation, that we (i.e. the inhabitants of the Earth) are in the absolute frame, which may be very unlikely, unless we accept the special relativistic view that effects like time dilation occur in each frame in compliance

---

the physical reality should not depend on our conventions, coordinate systems and frames.

[137] We should notice that the claimed effect of "aging less" or "returning younger" that the traveling twin will endure can only happen if we agree on the proposal that the "reality" and "truth" belong to the absolute frame; otherwise the traveling twin will see himself aging apparently as if he stayed at rest in the absolute frame. This may be a justification to our "convention" that the "reality" and "truth" belong to the absolute frame while all relative frames will have "apparent" reality and truth. Accordingly, this convention has real physical content if we accept the claimed effect of "aging less".

11.2 Essential Elements of Potentially Acceptable Interpretation 237

with the unrestricted relativity principle. What is your answer to this challenge?
Answer: Regarding experimental observations, like prolongation of lifetime of elementary particles, which may be claimed to support the special relativity view since it is very unlikely that the frame of the Earth is the absolute frame, it can be addressed by claiming that although the frame of the Earth is not the absolute frame, it is very likely that its speed relative to the absolute frame is a negligible fraction of the speed of light and hence for all practical purposes it can be assumed to be at rest with respect to the absolute frame when the speed of the observed phenomenon (i.e. the decaying elementary particles in our example) is comparable to the speed of light (i.e. $v \sim c$) or at least a considerable fraction of the speed of light. So, any deviation of our observations from the observations of the absolute frame is comparable in size to the experimental errors and hence it can be absorbed within these errors. In fact, this approximate identification of the frame of the Earth with the absolute frame should be very realistic (astronomically and cosmologically) if the absolute frame is identified by and based upon the spacetime structure as determined by the surrounding distribution of matter and energy. Also, this answer should apply to any similar challenge that is based on observing characteristic Lorentzian effects from the Earth (or indeed any ordinary frame of observation like spaceships where the speeds are very tiny fraction of c) as if it is the absolute frame.

10. Apply the proposed interpretation to a specific example, e.g. the prolongation of lifetime of elementary particles.
Answer: Let assume that the moving elementary particles are observed from the absolute frame (or from the Earth which is supposed to resemble the absolute frame according to the answer of the previous question). So, the time intervals and lengths in the frame of particles will be seen as contracted with respect to their values in the absolute frame. So, if we take their values in their proper frame as the inherent values then this is equivalent to having these values prolonged in the absolute frame in proportion to the contraction factor.[138] Hence, the lifetime as seen in the absolute frame will be prolonged in proportion to the contraction factor to obtain the observed lifetime in the absolute frame, and therefore the particles will traverse a prolonged distance in the absolute frame in proportion to their speed and lifetime. Alternatively, the traversed distance as seen in the absolute frame will be prolonged in proportion to the contraction factor to obtain the observed traversed distance in the absolute frame, and therefore the particles will have a prolonged lifetime in proportion to their speed and traversed distance in the absolute frame.

11. Can you provide more clarification about the relation between the velocity component of the light signal from its source and the invariance of the speed of light across all inertial frames?
Answer: Because the light signal obtains a velocity component from its source, its

---

[138] We should remark that because almost all proper values (including lifetimes) are measured on the Earth, which approximates the absolute frame, then the reality of this frame will justify the claim of being inherent values, and hence rationalize the reality of prolongation and add more sense to this explanation.

11.2 Essential Elements of Potentially Acceptable Interpretation

speed will be the same in any inertial frame that is in relative motion with respect to the source frame (refer to previous exercises as well as § 9.7.1), and because of this any frame other than the source frame will be like a source frame since it is not distinguished from the source frame by having a light signal whose speed is not c. For example, if we have three inertial frames (A, B and C) and the source of light is in one of these frames, say A, then when we transform this signal to B the speed in B will be c, so if we transform this signal (whether from A or from B) to C then the speed in C should also be c because there is no distinction between A and B in having the same speed c of the light signal whereas any difference in the relative motion between C and any one of A and B will be compensated by the spacetime contraction where this contraction is different for the two frames, A and B, if they have different relative speeds with respect to C. This means that although B is not physically the source frame, it behaves like A as a source frame. We think this is the culprit of the whole issue of the invariance of the speed of light across all inertial frames where this added velocity component from the source conspire with the spacetime contraction to ensure the invariance of the speed of light across all inertial frames and hence any frame can be regarded as a source frame for any light signal which means the speed of light in this frame is c regardless of the frame of the actual physical source of light (e.g. light bulb).

12. Let assume that the experimental evidence is consistent with the special relativity interpretation (e.g. in correlating the characteristic Lorentzian effects to the relative motion between frames in accord with the unrestricted relativity principle) but not with other interpretations (such as the proposed interpretation). What should we conclude?
Answer: In this case, Lorentz mechanics itself should be a wrong (or illusory) theory because it requires a logically inconsistent interpretation and hence all the observations that support Lorentz mechanics should be treated as empirical facts and therefore we should search for a consistent new theory (i.e. new formalism that can sustain a logically consistent interpretation) to justify these empirical facts. In brief, any formalism that requires logically inconsistent interpretation should be rejected (with its interpretation) as a valid scientific theory.

Chapter 12
Appendices

12.1 Maxwell's Equations

12.2 Michelson-Morley Experiment

1. What is the main limitation of the Michelson-Morley experiment and its analysis?
   Answer: The main limitation is that it is based on a classical wave model for the propagation of light since its main objective is to detect and measure the ether wind. Hence, even if the experiment is technically correct, it is not conclusive if there is a possibility that the light follows a projectile (or ballistic) propagation model because the source of light in the Michelson-Morley apparatus remains in a state of rest relative to the observer despite the rotation. In brief, the Michelson-Morley experiment may rule out the possibility of a classical wave propagation model[139] but it cannot rule out the possibility of a projectile propagation model for example. Accordingly, the Galilean transformations cannot be dismissed by the result of the Michelson-Morley experiment. We should remark that there are other limitations of the Michelson-Morley experiment (and its alike); some of these limitations have been indicated earlier.

12.3 Invariance of Laws under Galilean and Lorentz Transformations

1. What we mean when we describe a law as being invariant under certain transformations, e.g. Lorentz transformations?
   Answer: We mean that the law will take the same form when we move from one inertial frame to another inertial frame by applying a certain set of spacetime coordinate transformations. For example, the electromagnetic wave equation is invariant under the Lorentz spacetime coordinate transformations because it keeps its form when we change the coordinates in this equation by applying these transformations. But this equation is not invariant under the Galilean transformations of space coordinates and time because the form of this equation changes when we apply these transformations. The reader is referred to § 12.3.3 for more details about this example, and to § 1.8 and the exercises of § 6.1 for more details about form invariance.

2. Summarize what is invariant and what is not invariant in classical and Lorentz mechanics.
   Answer: According to classical mechanics, all inertial frames of reference are equally valid for formulating the laws of mechanics. Hence, all the laws of classical mechanics are invariant under the classical transformations of space coordinates and time which

---
[139] In fact, we already ruled out such a propagation model according to our analysis of the formalism of Lorentz mechanics, as explained earlier (see for example § 11).

are the Galilean transformations. However, the laws of electromagnetism, as represented by Maxwell's equations and their derived formulations, are not invariant under the Galilean transformations.

According to Lorentz mechanics, all inertial frames of reference are equally valid for formulating the laws of mechanics and electromagnetism. Hence, all the laws of mechanics and electromagnetism are invariant under the Lorentz spacetime coordinate transformations.[140]

12.3.1 Laws of Classical Mechanics

12.3.2 Maxwell's Equations

12.3.3 Electromagnetic Wave Equation

12.4 Derivation of Lorentz Spacetime Coordinate Transformations

12.4.1 Special Relativity Method of Derivation

12.4.2 Our Method of Derivation

1. Discuss the issue of lack of clarity and rigor in most (if not all) the derivation methods of Lorentz transformations.
   Answer: To appreciate the lack of clarity and rigor in these methods and their alike, the reader is advised to compare these derivations with the derivations of the velocity transformations (for example) which were presented in § 5.1.2. It should be obvious that while the latter are based on rigorous and obvious physical, mathematical and logical rules, the former are mixed with many hidden assumptions, arguable claims and hazy logic (among many other things). A main factor of this lack of clarity and rigor is their interpretative nature and that is why we prefer to postulate the Lorentz transformations in the first place instead of deriving them from other postulates. In fact, we can claim that some of these alleged derivations are no more than a smoke screen to pass certain interpretative (philosophical and epistemological) views and theories and market them as physically and mathematically rigorous formalism.

---

[140] In fact, the principle of form invariance in Lorentz mechanics should include all the laws of physics (within the validity domain of this mechanics). Moreover, we believe that this general invariance principle was formulated by some classical physicists even before the emergence of Lorentz mechanics (and it may be one of the causes of its emergence) and hence in this sense it can be considered as part of classical physics (although the classical transformations failed to satisfy this principle with regard to electromagnetism), as indicated earlier.

Index

3-operator, 179
3D space, 7–9, 13, 18, 97, 181, 182, 209
4-operator, 179
4-vector, 184–187
4D spacetime, 7–9, 12, 13, 89, 90, 95, 179, 181, 182, 209

Absolute
    derivative, 7, 184
        frame, 34, 41, 42, 46, 50, 54, 58, 66, 74, 76, 79, 80, 83–85, 140, 195–197, 202–204, 206, 208–212, 214, 223–225, 228, 231–237
        space, 34, 46, 49, 50, 52–54, 58–60, 66, 67, 76, 80, 84, 85, 202, 205, 210, 231, 234
        time, 12, 49, 51, 54, 73, 80, 194, 195, 202, 212, 214, 234
Abstract device, 13, 212
Accelerated motion, 36, 37, 59, 74, 80, 82, 83, 187, 224, 232, 233, 235
Accelerating frame, 50, 52–54, 66, 81, 83, 204, 224, 231
Acceleration, 34–38, 116, 140, 141, 164, 165, 184, 185
Anisotropy, 50, 76
Anti-identical, 100, 101, 196
Arc length, 8
Area, 105, 106, 108
Aristotelian philosophy, 48
Aristotle, 48
Atomic clock experiment, 200

Ballistic propagation model, 63, 209, 221, 239
Barn-pole paradox, 223, 225–227, 236
beta particle, 33
Blue shift, 142, 173, 174

Calibration, 74, 75, 96–98, 102, 111, 133, 159, 197, 206, 208, 221, 222, 232, 234, 235
Cartesian coordinate system, 9, 18, 73, 78, 123, 178–180, 185
Causal relation, 84, 85, 87, 88, 92–95, 100
Cause, 22, 53, 61, 64, 79, 94, 213
Characteristic speed, 7, 14, 16, 44–46, 56, 85, 86, 232
Charge density, 9, 146
Christoffel symbol, 9, 184, 185
Circular
    argument, 38, 42, 73, 79
    shape, 86, 87, 96, 143
Circularity, 169
Classical
    mechanics, 13, 15, 20, 26–28, 31, 33–35, 38–40, 48, 49, 51–54, 158
    view, 48–50, 55, 66, 217
Co-positional, 12, 13, 84, 85, 93, 99–101, 110, 111, 126, 127, 168, 191, 195, 196
Communication signal, 84, 95
Conservation
    laws, 154, 177
    of energy, 43, 44, 154, 155
    of energy-momentum, 186
    of kinetic energy, 30, 32, 176
    of Lorentzian mass, 154
    of mass, 23, 43, 44, 148, 154, 155
    of mass-energy, 154, 155
    of momentum, 24, 25, 29–32, 41, 43, 83, 104, 119, 150, 156, 168, 170, 177
    of total energy, 25, 119, 148, 149, 154–156, 176, 177
    principles, 177
Contravariant tensor, 7, 178, 179, 184
Coordinate system, 13, 18, 72, 73, 76, 78, 90, 97, 101, 123, 170, 179–182, 184, 187, 232, 236
Coulomb law, 84
Covariant
    derivative, 180
    tensor, 7, 178, 181, 182, 184
Curl, 7
Current density, 8, 146, 169
Curvature of spacetime, 50
Curved space, 70, 97
Curvilinear coordinate system, 184, 187
Cylindrical coordinate system, 9, 181

d'Alembert operator, 179
d'Alembertian operator, 7, 179, 180
Determinism, 21, 51, 52
Diagonal, 7, 97, 105, 106, 108, 109, 178, 180, 182, 183
Diffraction, 59
Displacement, 123, 235
    vector, 27, 184
Distance, 8, 22, 27, 28, 77, 84, 98, 99, 102, 103, 110, 193, 194, 196, 198, 199, 206, 219, 237

Divergence, 7, 179
Doppler shift, 145, 173, 174
Dragging coefficient, 8, 130
Dual field strength tensor, 8
Dynamic, 14

Earth, 37, 59, 61, 65, 66, 83, 88, 110, 113, 143, 216, 236, 237
Eddington, 11, 68
Einstein, 11, 12, 67, 68
Electric field, 7, 50
Electromagnetic
    scalar potential, 9
    vector potential, 7
Electromagnetism, 24, 65, 69, 124, 240
Electron, 8, 114, 118, 120, 148, 150, 152, 155, 159, 160
Elementary particle, 112, 199, 236, 237
Elsewhere region, 94, 98, 99
Energy, 118–122, 147–154, 171–175, 197
Epistemological, 10, 21, 22, 43, 71, 72, 79, 93, 97, 162, 211, 228, 230, 240
Ether, 11, 34, 46, 54, 55, 58–62, 64–67, 76, 85, 131, 202, 205, 210, 228
    drag, 61, 62, 65, 209
    wind, 59, 61, 64, 65, 67, 88, 89, 123, 239
Euclidean, 48, 92, 103, 182, 183
Event, 8, 12, 13, 18, 19
    space, 78, 93, 94, 98
Extrinsic property, 22, 23, 34, 52, 113, 114, 118, 141

Field strength tensor, 8
First postulate of special relativity, 224
FitzGerald, 11, 12, 64, 67
FitzGerald-Lorentz length contraction, 12, 61, 63, 64, 67, 68
Fizeau, 130, 131
    experiment, 131
    formula, 130, 131
Flat space, 70, 92, 94, 97, 180, 181
Force, 34–38, 117, 118, 171, 187, 188
Form invariance, 24, 25, 30, 31, 56–58, 66, 104, 178, 180, 239, 240
Frame of reference, 12–14, 16, 18–20, 77–84
Free
    particle, 92, 181, 187
    space, 9, 14, 16, 46, 47, 55, 59, 66, 71, 86, 130, 131, 133, 203
Frequency, 9, 45, 60, 141, 144, 145, 173, 174
    shift, 145, 173, 174
Fresnel, 130, 131
    drift effect, 131

Future, 13, 48, 50, 93, 94, 98

Galilean
    invariance, 53
    invariant, 30
    relativity, 52–54, 66
    transformations, 20, 24, 26–30, 32–35, 37, 38, 42, 47, 54–58, 61–63, 65–67, 75, 89, 101–104, 115, 116, 123, 124, 239, 240
Galileo, 55
General
    relativistic mechanics, 10
    relativity, 10, 11, 20, 43, 50, 68–70, 162, 203, 224, 231
Geodesic, 92, 181, 185, 187
Global
    observer, 73, 192, 213–215
    property, 192
Gradient, 7, 179
Gravitation, 37, 50, 69, 70, 84, 216, 218
Gravity, 10, 50, 70

Homogeneity, 50
Homogeneous coordinates, 180
Hooke law, 37

Identical, 84, 85, 99, 100, 195
Imaginary, 7, 13, 16, 86, 97, 152, 181
Improper
    length, 23, 64, 109, 190
    mass, 114, 170
    time, 168
    time interval, 109, 112, 167, 191
Inertia, 39, 41, 79
Inertial
    frame, 8, 10, 12, 14, 16, 18, 19, 24, 25, 27, 29, 30, 32, 34, 35, 37–39, 42, 50, 52, 53, 55, 58, 66, 67, 69, 73–75, 79–81, 83–86, 94, 95, 97, 98, 100–102, 104, 106, 111, 116, 123–126, 128, 130, 133, 134, 140–142, 144, 146, 159, 162, 166, 176, 177, 186, 196, 197, 202–211, 214, 222–224, 229, 231, 233, 235–239
    observer, 23, 25, 28–30, 42, 49, 83, 85, 101, 105, 107, 109, 112, 116, 126–129, 131, 132, 139–141, 143, 144, 166, 167, 193–196, 202–204, 206, 207, 215, 216, 224, 236
Inhomogeneity, 50, 76
Inner product, 183–185
Instantaneous rest frame, 143
Interference, 59, 88
    pattern, 67
Intrinsic

derivative, 7, 184, 187
property, 22, 38, 114, 118, 141
Invariance of
conservation of kinetic energy, 30, 176
conservation of momentum, 25, 29, 30, 104, 177
conservation of total energy, 176, 177
conservation principles, 177
electromagnetic wave equation, 104
kinetic energy, 176
laws of electromagnetism, 124
laws of mechanics, 57, 124
length, 104
light cone, 95
mass, 25, 39, 104, 168
Maxwell equations, 12, 56, 57
Newton's second law, 104
non-kinetic energy, 31
non-rest mass, 168
physical laws, 23, 24, 30, 101, 104
physical quantities, 104
quadratic form, 182, 183
rest mass, 168
space interval, 89, 104, 124, 164
spacetime interval, 89, 95–98, 101, 102, 104, 111, 124, 164, 182, 190, 197, 207–209, 215
speed of light, 58, 63, 75, 97, 98, 102, 111, 133, 134, 159, 162, 197, 202, 209, 210, 213, 216, 229, 231, 234–238
time interval, 89, 104, 124, 164
type of spacetime interval, 98
velocity of light, 231
Inverse
Lorentz matrix, 8
Lorentz tensor, 8
Isotropy, 50

Kennedy-Thorndike experiment, 65, 66
Kinematic, 36
Kinetic energy, 23, 29, 31, 32, 118, 119, 121, 122, 150, 152–155, 172, 176, 234

Laplacian operator, 7, 179
Larmor, 11, 12
Lateral speed, 86
Laue, 11, 67
Law of inertia, 39, 79
Length, 105–108, 166, 190
contraction, 12, 21, 23, 61, 63–67, 102–110, 112, 163, 166–168, 190, 191, 197, 199, 204–206, 210, 211, 219, 220, 223, 225–227, 229, 231, 232, 236
dilation, 167, 233

Light
clock, 212, 216–222
cone, 87, 93–96, 99–101
Lightlike, 94, 98–100, 102, 183
Line element, 7, 96, 97, 101, 180, 181
Local
observer, 193, 213–215
property, 192
time, 12
Logically
consistent, 22, 51, 85, 195, 205, 220, 224, 229, 232, 238
inconsistent, 205, 224, 238
Lorentz, 11, 12, 64, 65, 67, 68
factor, 9, 14, 15, 17, 65, 72, 86, 109, 162, 164, 171, 190, 191, 199, 223, 234
interpretation, 228
invariant, 176, 184–186, 207
matrix, 8
mechanics, 10–12, 64, 66–72, 89
tensor, 8
transformations, 65, 125, 127, 163, 164, 239, 240
Lorentzian mass, 148, 154, 172
Luminiferous
ether, 55, 58–60, 66
medium, 59

Mach principle, 41
Magnetic field, 7, 50
Manifold, 12, 13, 48, 70, 89, 90, 92, 96, 97, 101, 103, 104, 111, 178, 179, 181, 190
Mass, 114, 141, 147–150, 168, 169, 173, 174, 197
Mass-energy
equivalence, 118, 120, 148, 149, 197
relation, 118, 148, 172–174
Massive, 12, 16, 24, 29, 30, 39, 41, 47, 86, 92, 96, 99, 114, 115, 118, 121, 139, 141, 151, 152, 158–162, 172, 187, 202, 216, 218, 235
Massless, 12, 16, 86, 88, 151, 152, 159–162, 202, 235
Maxwell, 66
equations, 12, 54–59, 61, 65–67, 123, 124, 240
Metric tensor, 7, 97, 178–183
Michelson, 63
Michelson-Morley
analysis, 88, 217
apparatus, 67, 239
device, 65
experiment, 12, 58, 59, 61–67, 88, 123, 209, 217, 218, 222, 239
method, 62, 65
Minkowski, 11, 67, 68, 211

metric tensor, 185
space, 12, 89, 90, 96, 97
spacetime, 7, 12, 92, 94, 179–183, 185, 187, 209
Mixed tensor, 179
Modern convention about mass, 39, 104, 168, 186
Momentum, 116, 117, 151–153, 169, 170, 175, 185, 186
Momentum-energy relation, 152, 175, 186
Morley, 63
Mutually exclusive, 94, 99, 100

nabla operator, 7
Neutron, 8, 117, 120, 122, 148, 152
Newton, 49
first law, 23, 39, 40, 42, 81, 83, 118, 141, 147, 181, 187
law of gravity, 84
laws, 41, 42, 47, 79, 80, 82, 83, 147
second law, 24, 35, 38–43, 82, 104, 117, 147, 170–172, 177, 187
third law, 40, 42, 82, 147
Newtonian principle of relativity, 53
Non-
accelerating, 52
Cartesian, 78
inertial frame, 14, 23, 42, 52–54, 75, 79, 80, 84, 86, 141, 162, 203, 204, 209–211, 214, 231, 233, 235
inertial observer, 23, 83, 143
invariance of Maxwell equations, 56
invariance of space interval, 164
invariance of time interval, 164
invariance of velocity of light, 231
kinetic energy, 31, 50, 118, 154, 172, 176
local reality, 84, 88
rest mass, 39, 114, 168, 169, 172
simultaneous, 192, 195, 212, 214, 215
Null
cone, 94
geodesic, 92, 94, 100

Observed speed, 16, 45, 46, 50, 56, 59, 85, 202, 203, 207, 208, 218, 219
Observer, 15, 17
Old convention about mass, 39, 168, 172, 186

Partial derivative, 7, 180
Past, 13, 48, 50, 92–94, 98
Permeability of free space, 9, 55, 59
Permittivity of free space, 9, 55, 59
Phase shift, 62, 65
Philosophical, 10, 21, 22, 43, 48, 67, 73, 84, 93, 97, 162, 211, 228, 230, 240

Physical reality and truth, 21, 205, 210, 229
Planck, 11, 67
constant, 174
relation, 174
Poincare, 11, 12, 67, 68
mass-energy relation, 118, 148, 154, 155, 172
synchronization procedure, 75–78
Position vector, 8, 184
Positive definite, 97, 182
Postulate of
constancy of speed of light, 162
invariance of speed of light, 202
relativity, 162, 202
Postulates of special relativity, 11, 163, 203
Preciseness, 21, 51, 52
Present, 13, 48, 50, 93, 94, 98
Principles of reality and truth, 21–23, 95, 205, 209, 223, 229, 232
Projectile propagation model, 63, 66, 75–77, 88, 89, 133, 134, 209, 210, 213, 217, 221, 222, 232, 239
Prolongation of lifetime, 236, 237
Propagation medium, 46, 47, 50, 58, 85, 210, 222
Proper
area, 105
charge density, 9
frame, 237
frequency, 9, 144
length, 7, 8, 64, 105, 106, 109, 190, 217, 223, 225, 226, 234, 236
mass, 141
time, 8, 179, 191, 217, 223
time interval, 109, 112, 127, 128, 167, 191
time parameter, 9
volume, 8
wavelength, 9, 143
Proper-improper perspective, 102, 109, 191
Proton, 8, 12, 116, 119, 120, 148, 150, 152
Pythagoras theorem, 220

Quadratic
form, 180–183
formula, 139, 153

Red shift, 142, 143, 173, 174
Reflection, 13, 94, 232
Relativity
of co-positionality, 126, 196, 197, 234
of simultaneity, 126, 192–197, 212–215, 234
Rest
energy, 118, 120, 121, 148, 150, 152, 155, 172, 176, 185

INDEX

245

frame, 12, 33, 44–46, 52, 55, 59, 60, 66, 74, 76, 107, 108, 112, 128, 129, 134, 141, 146, 166–168, 179, 190, 199, 209, 213, 217, 219, 223
mass, 8, 39, 104, 114, 118, 154, 168, 172, 176, 185, 186

Restricted
relativity principle, 204
speed, 16, 71, 86, 162, 202, 235

Rotation, 13, 14, 18, 20, 54, 59, 64, 80, 143, 184, 218–220, 232, 239

Scaling, 13, 49, 112, 206, 208, 232

Second postulate of special relativity, 17, 75, 77, 95, 97, 98, 203, 209, 213, 218–220

Shearing, 13

Simultaneity
of observation, 192, 193, 195, 196, 212–215, 234
of occurrence, 192, 193, 195, 197, 212–215

Simultaneous, 12, 13, 84, 99, 100, 112, 126, 168, 190–195, 212–215

Singular, 152

Singularity, 16, 86

Solar eclipse expedition, 11, 68

Sommerfeld, 11

Space, 26–28, 72, 73, 190
coordination, 72, 73
interval, 8, 12, 13, 97, 104, 124, 164, 190, 222

Spacelike, 94, 98–100, 183

Spacetime, 89–92, 94–104, 183, 190
contraction, 133, 166, 184, 190, 191, 195, 197, 206, 210, 213, 214, 222, 231–233, 235, 236, 238
diagram, 90, 92, 192, 193, 209
distortion, 133, 197
interval, 9, 12, 92, 95–101, 103, 104, 111, 124, 164, 180, 182, 183, 190, 197, 207–209, 215, 222

Spatial separation, 28, 101, 103, 112, 126, 127, 163, 168, 193, 195

Special
relativistic mechanics, 10
relativity, 10, 11, 16, 21, 42, 43, 47, 53, 61, 62, 67–69, 71, 75, 77, 85–88, 95, 97, 98, 162, 195, 202–205, 209, 211–214, 216–220, 223–225, 227, 228, 230, 231, 235, 237, 238

Speed, 45, 157–161
of light, 7, 46, 47, 55, 56, 85–89, 206–209
of projectile, 44, 45
of wave, 44, 45
ratio, 9, 14, 15, 17

Spherical coordinate system, 8, 182

Standard setting, 18–20, 25–30, 34, 35, 37, 38, 102, 103, 106, 111, 123–126, 128–134, 140, 159, 163, 166, 168, 186, 191, 193–196, 206, 207, 211, 214, 215, 236

Summation convention, 17, 99, 178, 181

Sun, 59, 65, 213

Temporal separation, 163, 168, 195

Thought experiment, 42, 43, 212–215

Time, 26–28, 73–77, 89, 109–113, 166–168, 190, 191
contraction, 167, 191, 233, 236
dilation, 12, 21, 23, 61, 65, 74, 91, 102–104, 109–111, 113, 128, 163, 166–168, 175, 176, 190, 191, 197–200, 204–206, 210, 211, 214, 216–220, 223, 224, 229, 231–233, 236
dilation triangle, 175
interval, 20, 27, 28, 36, 49, 101, 104, 109, 110, 112, 124, 126, 127, 163, 164, 166, 167, 190, 191, 193, 195, 206, 221, 222, 237
synchronization, 73–76

Timelike, 94, 98–101, 183

Total
derivative, 187
energy, 7, 25, 118, 119, 121, 148, 149, 151, 152, 154–156, 172, 175–177

Train thought experiment, 43, 212, 213, 215

Trajectory, 12, 72, 91, 92, 117, 181, 216, 218

Transformation of
acceleration, 38, 140, 141, 164
length, 166
mass, 141
space, 28, 29, 35, 54, 56, 58, 69, 157, 167, 168, 190, 197, 211, 239
spacetime, 65, 67, 70, 97, 101, 103, 111, 112, 114, 116, 123, 124, 151, 160, 162, 163, 180, 190, 202, 206, 207, 213, 227, 239, 240
time, 28, 29, 35, 54, 56, 58, 69, 103, 157, 164, 167, 168, 191, 192, 194, 196, 197, 214, 221, 239
time interval, 166
velocity, 29–32, 38, 70, 115, 128–132, 134, 151, 157, 160, 164, 207, 209, 240

Translation, 13, 14, 18, 20, 25, 40, 41, 49, 54, 59, 79, 143, 232

Twin paradox, 205, 211, 223, 224, 236

Ultimate speed, 16, 71, 87, 95, 162, 202, 235

Uniform
acceleration, 13, 14, 20, 70
motion, 12, 23, 27, 28, 33, 41, 42, 52, 53, 60, 66, 80, 83, 109, 181, 195, 203, 204, 209, 231–233, 235

INDEX

Unrestricted relativity principle, 204, 237, 238

Vacuum, 7, 58
Value invariance, 24, 25, 30, 31, 39, 104
Velocity, 28–34, 114, 115, 128–132, 134–137, 139, 140, 164, 184
    composition, 33, 34, 54, 58, 61, 85, 88, 134–137, 139, 145, 146, 157, 160, 161, 218, 219, 221
Voigt, 11
Volume, 8, 107, 108

Wave propagation model, 63, 66, 75–77, 88, 89, 133, 210, 212, 217, 221, 239
Wavelength, 9, 141–143
Work, 122, 148, 154
Work-energy relation, 154
World line, 12, 78, 90–92, 96, 187